Design and Implementation of Real-Time
Multi-Sensor Vision Systems

Vladan Popovic • Kerem Seyid • Ömer Cogal
Abdulkadir Akin • Yusuf Leblebici

Design and Implementation of Real-Time Multi-Sensor Vision Systems

Vladan Popovic
Intel Corp
Santa Clara, CA, USA

Ömer Cogal
Heptagon AMS Group
Rüschlikon, Switzerland

Yusuf Leblebici
EPFL Microelectronic Systems
  Laboratory
Lausanne, Switzerland

Kerem Seyid
Lausanne, Switzerland

Abdulkadir Akin
Quantum Engineering Center
ETH Zurich
Zürich, Switzerland

ISBN 978-3-319-86539-3   ISBN 978-3-319-59057-8   (eBook)
DOI 10.1007/978-3-319-59057-8

This Springer imprint is published by Springer Nature
The registered company is Springer International Publishing AG
The registered company address is: Gewerbestrasse 11, 6330 Cham, Switzerland

Preface

The Microelectronic Systems Laboratory (LSM) of EPFL first started working on multi-camera vision systems back in 2009, when the concept of omnidirectional imaging using multiple sensors was still in its infancy. At that time, our research group took up the problem of building multi-sensor real-time imaging systems mainly as a challenge that required innovative solutions combining algorithms, system architectures, and hardware implementations. Conventional graphics processing hardware architectures prove remarkably inadequate for a number of tasks required for multi-sensor imaging. This called for a wide range of custom solutions that sought to optimize pixel-level processing, memory access, and parallelization. When we started our research activities in this domain, there were not even any user applications, nor any system specifications anticipated for multi-sensor vision systems.

Fast-forward to 2017, we see a very active field and a proliferation of exciting applications of multi-camera systems in virtual reality (VR), augmented reality (AR), security, surveillance, and many other areas. The research carried out by members of LSM during the past 7 years allows us to address a multitude of problems associated with multi-sensor imaging, including omnidirectional panorama construction, image blending, networks of cameras, disparity estimation—among others. Throughout the years, our research group has been fortunate to be at the forefront of this very exciting and fast-developing domain.

This book covers a number of topics that have been explored in the context of multi-camera vision systems, discussing aspects related to specialized algorithms as well as the associated hardware design and implementation. As such, the book offers a comprehensive theoretical basis for the understanding of multi-camera systems. We cover the basics of image formation, algorithms for stitching a panoramic image from multiple cameras, and multiple real-time hardware system architectures, in order to obtain panoramic videos. The topics covered in this book would be beneficial to graduate students specializing in omnidirectional multi-sensor systems, as well as HW/SW engineering professionals designing future systems for a wide range of VR/AR applications.

The authors are truly indebted to many individuals who have contributed to this work over the years. The early phases of this research were initiated by the very fruitful collaboration with Professor Pierre Vandergheynst and his group, who first brought the systems design challenge to the attention of the LSM team. Dr. Alexandre Schmid and Dr. Hossein Afshari of LSM have been instrumental in establishing the theoretical basis for the first multi-camera systems explored in this work, and in designing the first prototypes. Over the years, Mr. Sylvain Hauser and Mr. Peter Brühlmeier have provided their excellent skills for building some of the most unique and most complicated hardware platforms that enabled the experimental work presented here. The strong support and valuable advice offered by our colleagues at Armasuisse, Dr. Peter Wellig and Dr. Beat Ott, are highly appreciated. We thank our graduate project students, in particular, Mr. Jonathan Narinx, Mr. Selman Ergunay, Mr. Bilal Demir, Mr. Luis Gaemperle, Ms. Elieva Pignat, Ms. Ipek Baz, Mr. Andrea Richaud, and Mr. Raffaele Cappocia, for their contributions. We acknowledge the very valuable and consistent support of Mr. Bentley Woudstra and Mr. Karl Osen, who have successfully motivated the effort since the very early days. Finally, a very special note of thanks goes to Mr. François Stieger for his extraordinary support and encouragement. This work would not have been possible without their contributions.

Santa Clara, CA, USA   Vladan Popovic
Lausanne, Switzerland   Kerem Seyid
Rüschlikon, Switzerland   Ömer Cogal
Zürich, Switzerland   Abdulkadir Akin
Lausanne, Switzerland   Yusuf Leblebici

Contents

Chapter 1
Introduction

For more than a century, photographs were taken by exploiting the chemical reaction of light rays hitting the photographic film. The photographic film is a plastic film that is coated with light-sensitive *silver-halide* crystals. The emulsion will gradually darken when exposed to light, but the process is too slow and incomplete to be of any practical use. A principle of *camera obscura* is used to resolve this problem. The *camera obscura* ("dark room"—Latin) is an optical device consisting of a box and a small hole (pinhole) on one of its sides, as shown in Fig. 1.1a. Light that passes through the hole projects an inverted image of the world on the opposite side, but with preserved color and perspective.

Even though the first written records of *camera obscura* can be found in the fourth century BC in the works of Chinese philosopher Mozi, and the Greek philosopher Aristotle, the first clear description and extensive experiments were published by Leonardo da Vinci in 1502. The modern cameras are based on this principle, but do not use the pinhole as it produces dim or blurry images, depending on the hole size. A lens, or a system of lenses, is used to focus the light, producing the usable brightness levels and sharp images. The image is projected on the photographic film placed on the opposite side of the lens. A short exposure to the image formed by a camera lens is used to produce only a very slight chemical change, proportional to the amount of light absorbed by each crystal. This creates an invisible latent image in the emulsion, which can be chemically developed into a visible photograph.

It was more than 200 years until the first commercial camera was produced. The "Giroux Daguerreotype," in Fig. 1.1b, was presented in Paris in 1839. It was considered to be a very fast camera, with the exposure time of around 3 min, depending on the scene illumination. The oldest surviving photograph is shown in Fig. 1.2. It was shot with one of the test prototypes of Giroux's camera.

Technological advancements in the late twentieth and early twenty-first century also affected the camera manufacturing process. The cameras moved from the analog capture process using photographic film to the digital one. The photographic film has been replaced by the digital image sensors that convert light, i.e., photons,

© Springer International Publishing AG 2017
V. Popovic et al., *Design and Implementation of Real-Time Multi-Sensor Vision Systems*, DOI 10.1007/978-3-319-59057-8_1

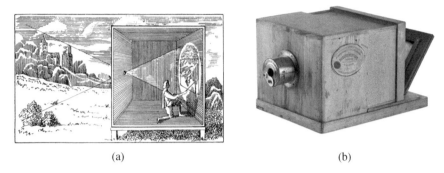

(a)   (b)

Fig. 1.1 (**a**) Illustrated principle of *camera obscura*, and (**b**) the first commercially available camera produced by Louis-Jacques Daguerre and Alphonse Giroux

Fig. 1.2 "View from the Window at Le Gras" is the oldest surviving photograph. It was taken in 1826 or 1827

into electrical signals. Two main types of digital sensors are used today: charge-coupled devices (CCD) and active pixel sensor in complementary metal-oxide-semiconductor (CMOS) technology. Most of the modern-day sensors use CMOS technology, thanks to the lower price, simpler design, and faster readout of pixels. The big boom in the camera use started about a decade ago, with the appearance of mobile phones with the integrated cameras, and later smartphones. The cameras are now massively produced, they are easily accessible, and they are becoming cheaper.

1.1 Computational Imaging

Research in the fields of image processing and computer vision is based on obtaining truthful representation of the visual world, and adapting the acquired data to the desired purpose. The digital image sensors are used to measure the light intensities in the real world. Based on these measurements, various algorithms have been developed to reduce image acquisition artifacts (noise, lens distortion, chromatic aberration, etc.), compress the size of the image without losing quality, or apply some of the advanced methods for understanding contents of the image (edge detection, image segmentation, object recognition, etc.).

With the increased availability of the image sensors, the researchers started looking at the new possibilities in the image acquisition domain. Conventional imaging technique is limited to representing the world in a two-dimensional (2D) frame. The need to truthfully represent the environment initiated changes in the way the images are taken. Computational imaging, or computational photography, is a new vibrant research field in the last decade. It differs from the conventional imaging by utilizing a mix of novel optics and computational algorithms to capture the image. The novel optics is used to map rays in the light field of the scene to pixels on the detector in some unconventional fashion [7].

The fundamental idea of a computational camera is to use a combination of optics and sensors to optically encode scene information. The image captured by such a device does not necessarily look like the scene we see with our eyes. Hence, computational cameras have the processing (computing) step, which translates the recorded data into a meaningful image that our brain can understand. These systems are used to provide new functionalities to the imaging system (depth estimation, digital refocusing, super-resolution, etc.), or to increase system's performance (image resolution, camera frame rate, color reproduction, etc.).

Figure 1.3 illustrates the most popular design methodologies of computational cameras. Object side coding in Fig. 1.3a is the easiest way to implement a computational camera. It denotes any kind of attachments external to the camera, such as adding a mirror (detailed in Sect. 2.2.2) or a prism. Figure 1.3b shows a focal plane coding method where an optical element is added close to the image sensor. The example of potential optical elements is a pixel filter for multispectral and high dynamic range imaging, or a microlens array used to capture the light field (Sect. 2.2.5). Figure 1.3c shows an example where the computational camera uses controlled illumination, such as the coded structured light [8]. Finally, Fig. 1.3d shows an omnidirectional (*omni*—"all" in Latin) imaging system that utilizes a multiple camera setup. These systems will be the main topic of this book, starting from the design constraints and ending with several potential applications.

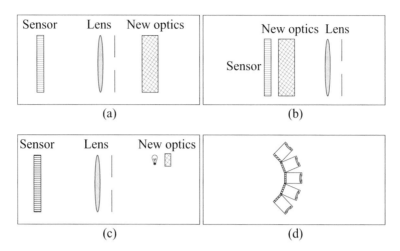

Fig. 1.3 Various design methodologies of computational cameras [7]: (**a**) Object side coding, (**b**) focal plane coding, (**c**) light coding, and (**d**) multi-camera systems

Fig. 1.4 The oldest surviving panoramic photograph. It shows the view of San Francisco in 1851 from Rincon Hill

1.2   Bridging the Performance Gap

Multi-camera systems with cameras arranged in a circle or a sphere are suitable for capturing panoramic images and videos. Panoramas were originally created by taking several photographs with a single camera, developing the film, and then manually stitching photographs one to another (Fig. 1.4). It was mostly used in artistic, or military purposes, such as wide area surveillance. Today, there is a plethora of algorithms for the automatic image stitching and panorama construction. The most used ones will be explained in Chap. 3.

A common pipeline of panorama construction is to take images using one or more cameras, order them, and stitch into a single wide field of view (FOV) image. The stitching process is usually done offline on a personal computer (PC). Such approach does not have timing constraints, and the applied image processing algorithms can be very intensive and time-demanding. As a result, the image quality of the panorama can be very high.

However, real-time operation is needed in applications such as medical imaging for diagnostics or during surgeries, surveillance, or autonomous vehicles. Currently, the real-time operation is possible only with a single (or very few) camera and with a limited resolution. If a wide FOV is required, the systems are designed with a *fish-eye* lens, or with a camera mounted on a movable stand [1].

This book focuses on the design of computational imaging systems operating in real-time and with high image resolution. We use the high performance processing platforms based on field-programmable-gate-array (FPGA), which allow faster computations and real-time operation. FPGA systems are suitable for this purpose since they are programmable, and provide fast prototyping compared to the design of a dedicated application-specific-integrated-circuit (ASIC). A complete system-on-chip, which includes a microprocessor and its peripherals, can be inferred in the FPGA. In order to achieve the highest possible speed, the dedicated hardware processing blocks are created. Hence, there is no image processing done on the microprocessor, resulting in significant increase in performance.

The currently existing panorama construction software algorithms cannot be efficiently ported to the hardware processing platforms. Hence, they need to be modified and adapted for fast processing on FPGA, without any loss of image quality compared to its software counterpart. Furthermore, we will present the design pipeline of a novel, modular and fully scalable multi-camera system, able to reach Gigapixel resolution, that still operates with a real-time frame rate of 30 frames-per-second (fps). Up to our knowledge, this is currently the fastest omnidirectional imaging system.

1.3   Miniaturized Panoramic Imaging

Miniaturization in general is the key trend for many years in different technological areas of interest. Recent increase in mobile, robotics, and medical applications of imaging and vision systems forced scientists and engineers to develop new miniaturization solutions of both electrical and mechanical aspects of cameras.

Recent developments of the drones bring new requirements for machine vision systems utilized in aerial applications [3]. Since such flying vehicles are light-weight, power-limited systems and require large FOV imaging, miniaturized panoramic imaging solutions are becoming crucial. Likewise, in the mobile phone and hand held-devices industry, there are many new systems attempted to add the panoramic imaging capabilities to the new generation mobile devices [9]. Again, the device size is becoming the key aspect in this area.

In addition to the mentioned applications, certain applications in the medical imaging field require large FOV. Colonoscopy and laparoscopic surgery are very well-defined examples of these applications. In laparoscopic or minimal invasive surgery (MIS), the need for miniaturized panoramic imagery is an inevitable requirement since it is crucial to have both miniaturized imaging and capability of large FOV to be able to see the whole surgery area. Current systems in MIS domain utilize relatively small field of view cameras and need movement of the imaging equipment during operation to see certain areas in the region of interest.

From the applications and needs from the real-world problems, the first question seeking for an immediate answer is how to have a wide FOV or (panoramic) imaging system, which has relatively high resolution in a limited volume that can fit into applications like flying drones, smart phones, or medical applications, such as colonoscopy and minimal invasive surgery.

1.4 Insect Eyes

The nature has its solutions for panoramic imaging. The compound eye structure of insects is a good example of this kind of solution. In Fig. 1.5, the structure of an insect eye is shown. The insect eyes, at their outer shell, have small lens facets adjacent to each other, looking into different directions. Hence, they have the 360° vision of their surrounding world. According to their type of vision, they have one or more sensors named as rhabdoms under each lens. Each unit with a lens and a sensor is called ommatidia. These units collect light from different angles with the help of their lenses, turn the absorbed light into a signal with the corresponding sensor, and interpret the information in their brain [5].

Insect eyes are divided into different categories according to the way they form the image. These types are apposition, optical super position, and neural superposition [4, 5]. In apposition type, the lens and the sensor pairs are separated from neighboring pairs forming a small portion (1–2°) of the full field of view. In optical superposition type, the light collected by adjacent lenses is focused and superposed optically on each of the neighboring sensors, creating a lumped

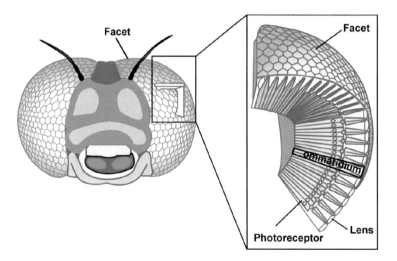

Fig. 1.5 Anatomy of the insect eye; image from [2]

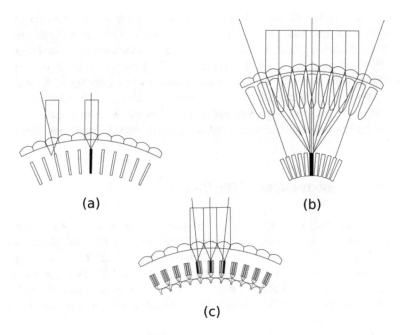

Fig. 1.6 Insect eye types according to their image formation, image redrawn from [4] (**a**) apposition, (**b**) optical superposition, (**c**) neural superposition type

information of the surrounding light field on each sensor. In neural superposition type, the sensor signals from the adjacent rhabdoms are superposed in the signal domain on the way to the brain or in the brain. The illustration is given in Fig. 1.6 for the different kinds of insect eyes.

1.4.1   Bio-Mimicking Problem of Insect Eyes

Many attempts were made to emulate the insect eyes in imaging systems for large FOV or panoramic imaging. The reasons for mimicking these natural vision systems are numerous. First of all, insect vision system provides a large FOV. When compared to a single wide FOV cameras, they provide more scalability and uniform resolution. Due to the distributed behavior, they provide fast information processing. However, their main drawback is the limited optical resolving capability, due to the limited size and optical effects such as diffraction. A review [6] states that blind one-to-one mimicking of the insect eyes is not the best approach. Thus, a system that is both small in dimension, has a high resolving capability, and large FOV is not an easily achievable design.

The mentioned trade-off items form a research space for the insect eye-mimicking problem. This is one of the driving motivations for this book; to search and find an optimal solution for having a miniaturized, high resolution, and large FOV imaging system. This image processing system includes both optical and electronic components to provide a real-time video capability (at least 25 fps) as well. Therefore, the main goal is to propose a set of algorithms and methods for designing and manufacturing insect eye inspired large FOV multi-aperture systems. Apart from that, we explore the size and the quality limits of the proposed methods.

1.5   Key Contributions of the Book

Each result presented in this book has originated from, and was motivated by a practical problem in the fields of computational imaging and embedded systems design. The problems at hand have defined the methods to be used in this book. Thus, the contributions of the book are presented by their corresponding applications, i.e., in each chapter we start by explaining the practical problem at hand, and then we present the methods to solve such a problem, together with their practical implementation.

The main aspects of this book can be classified in three categories: algorithm development for panorama construction, multi-camera system design, and novel camera applications. The contributions include for the first category:

- The panoramas are created by projecting the single camera images onto a surface, such as a sphere or a cylinder. The panoramas are then displayed by sampling the sphere using equiangular sampling. We provide **a novel constant pixel density sampling scheme**, which guarantees that each camera used in the system contributes to the panorama with approximately the same number of pixels. The advantage of this scheme is two-folded: the processing load is equalized among the cameras, and distortions in the spherical panoramas are reduced, thanks to reduced influence of the cameras observing the poles.
- Different methods are proposed for taking advantage of information from the overlapping FOV of multiple cameras, which is not possible to achieve by a single camera system. These methods can be listed briefly as: **utilizing parallax information as object edge features**, and **using the inter-camera intensity differences as evidence for generating a final compound panoramic image**.
- The formulation of the **novel image blending method named Gaussian blending**. This new method is based on a weighted average with the per-pixel-per-camera weights, as opposed to the standard per-camera weights. The Gaussian blending method resolves artifacts observed in the current state-of-the-art approaches, such as ghosting or visible seams. Furthermore, the Gaussian blending is **suitable for a fast real-time hardware implementation**.

The second category focuses on the hardware design of multi-camera systems as individual case studies of different architectures:

- The full design of the Panoptic camera that is **the first real-time omnidirectional multi-camera system**. It is also **the first computational imaging system to use an FPGA as the processing core**, which combined with the novel image processing architectures results in real-time **panoramic video** construction.
- The full design of an **interconnected network of smart cameras**. In this multi-camera system, each camera is provided with smart capabilities such as local memory storage, limited local image processing, and communication interface to other smart cameras in the network.
- The full design of a **miniaturized multi-camera system for endoscopy applications**. The analysis includes optimal camera placement, integrated illumination of the environment, mechanical miniaturization, and FPGA design for real-time image processing on such system.
- **The fully modular and scalable multi-camera system**, GigaEye II, designed for **gigapixel video** construction. The system uses multi-layered processing architecture with stacked FPGA processing boards. This system is able to process gigapixels of data in real-time and create a panoramic video output at high resolutions and high frame rates. Compared to the state of the art, GigaEye II is **the fastest multi-camera system**, and the whole processing pipeline is implemented in FPGA, hence it does not require a large cluster of computers for the image processing.
- **Binocular and trinocular** real-time depth estimation systems.

Finally, this book presents the new applications for the computational imaging systems:

- We present the first system targeted at **virtual reality applications** that includes both the acquisition device (Panoptic camera) and the virtual reality goggles as a display device (Oculus Rift).
- A new method to record and **reconstruct the high dynamic range video** using multiple cameras. We designed a multi-camera system able to utilize the overlap in the cameras' field of view to **recover the high dynamic range radiance map**, **stitch the panoramic frame**, and **apply the tone mapping** for proper display. All of the operations are done in real-time.
- Using our depth estimation system, we also present the hardware aimed for the **real-time free view synthesis** application.

1.6   Book Outline

This book is organized as follows. Chapter 2 gives an overview of the state-of-the-art computational imaging systems and panorama construction algorithms.

Chapter 3 introduces the panorama construction algorithms, together with their advantages and disadvantages. The image formation in a single camera is also explained, as well as the explanation of how we can use the camera geometry to correctly stitch the panorama. This chapter introduces illumination differences as the main problem in panorama generation. The major contributions of this chapter are the proposed Gaussian image blending solution that can be implemented in a real-time camera system.

The first case study of this book is given in Chap. 4, where the first real-time multi-camera system is presented. It is a miniaturized and easily portable system. The chapter describes the full design procedure, including the image acquisition and the real-time processing.

The second case study is given in Chap. 5, where we analyze and model the insect eye for miniaturization with pinhole cameras. The camera placement methods are revised and an optimization of the camera placement method for the target system is disclosed. The chapter also includes the image processing implementations for generating a seamless compound panoramic image, and the fabrication details of the miniature compound eye.

The third case study is given in Chap. 6. We present a novel distributed and parallel implementation of the real-time omnidirectional vision reconstruction algorithm. In this approach, new features are added to the camera modules such as image processing, direct memory access, and communication capabilities in order to distribute the omnidirectional image processing into the camera nodes. We also present a way to immersively display the panoramic video using a head-mounted display (HMD).

The fourth case study of this book is given in Chap. 7, which presents the very high-resolution imaging system called GigaEye II. This chapter also explains all considerations that have to be taken into account when designing such systems. Novel processing architectures and implementations are given in detail. Furthermore, the system is designed in a modular fashion, and new cameras can easily be added to increase the resolution even further.

The depth estimation algorithm and its hardware implementation is presented in Chap. 8. We present the novel binocular disparity estimation algorithms AWDE and AWDE-IR and their hardware implementations. The trinocular disparity estimation algorithm T-AWDE and its hardware implementation are also presented in the same chapter, together with the full embedded system for the disparity estimation.

In Chap. 9, we focus on designing an FPGA-based real-time implementation for correct image registration. We introduce a novel image registration algorithm based on the optical flow calculation. The presented hardware is using a hierarchical block matching-based optical flow algorithm designed for real-time hardware implementation.

Finally, a set of applications, where these computational imaging systems can be used, is given and demonstrated in Chap. 10.

Chapter 11 concludes this book and states ideas for the future research in this field.

This book is written as a follow-up of four PhD theses defended by the authors. Chapters 1, 2, and 11 were written by all authors. Chapters 4 and 7 were written by Vladan Popovic, Chaps. 6 and 9 by Kerem Seyid, Chap. 5 by Ömer Cogal, and Chap. 8 by Abdulkadir Akin. Chapter 3 was written jointly by Ömer Cogal and Vladan Popovic, and Chap. 10 by Abdulkadir Akin and Vladan Popovic.

References

1. Cogal O (2015) A miniaturized insect eye inspired multi-camera real-time panoramic imaging system. PhD thesis, Swiss Federal Institute of Technology (EPFL)
2. EPFL laboratory of intelligent systems- curvace (2015) http://lis.epfl.ch/curvace. Accessed 2015-09-06
3. Franceschini N (2014) Small brains, smart machines: from fly vision to robot vision and back again. Proc IEEE 102(5):751–781. doi:10.1109/JPROC.2014.2312916
4. Land MF (1997) Visual acuity in insects. Ann Rev Entomol 42(1):147–177
5. Land MF, Nilsson DE (2012) Animal eyes. Oxford University Press, Oxford
6. Lee LP, Szema R (2005) Inspirations from biological optics for advanced photonic systems. Science 310(5751):1148–1150
7. Nayar S (2011) Computational cameras: approaches, benefits and limits. Technical report, Computer Vision Laboratory, Columbia University
8. Salvi J, Pages J, Batlle J (2004) Pattern codification strategies in structured light systems. Pattern Recogn 37(4):827–849. doi:10.1016/j.patcog.2003.10.002
9. Thibault S, Parent J, Zhang H, Roulet P (2014) Design, fabrication and test of miniature plastic panomorph lenses with 180 field of view. In: International optical design conference. International Society for Optics and Photonics, p 92931N

Chapter 2
State-of-the-Art Multi-Camera Systems

This chapter presents both pioneering and state-of-the-art algorithms and systems for acquisition of images suitable for wide FOV imaging. The most common systems include translational and rotational single camera systems, catadioptric cameras, multi-camera systems, and finally, commercially available standard and light-field cameras. We also introduce the systems mimicking insect eye, and how camera systems can be used to estimate depth of the scene.

2.1 Panorama Stitching Algorithms

Panoramas have a much wider FOV than the standard cameras can provide. They are usually created by stitching several images taken by one or more cameras into a single one. The acquired images should have overlapping regions that are used for alignment of the original images. The creation of panoramas or image mosaics has been a popular research topic over the past years. The process of creating a panorama consists of various processing steps, which are named differently in the literature. For the sake of simplicity and consistency in this thesis, the process of stitching multiple images is divided into two main parts: (1) proper image alignment, and (2) seamless image blending.

The purpose of the image alignment is to determine the correct orientation and position of the original images in the final mosaic. Images taken by mobile devices are not horizontally aligned and often have slight pitch and roll rotations. Additionally, images taken by different cameras may have noticeable differences in scaling, due to varying fabrication process of the camera's optical system. Various algorithms for aligning the captured images were developed in the last two decades. Szeliski and Shum [85] presented a method to recover 3D rotations between the images. Brown and Lowe [11] use shift-invariant features to find matches between all images. Thanks to the matching between all images, this method is robust to

© Springer International Publishing AG 2017
V. Popovic et al., *Design and Implementation of Real-Time Multi-Sensor Vision Systems*, DOI 10.1007/978-3-319-59057-8_2

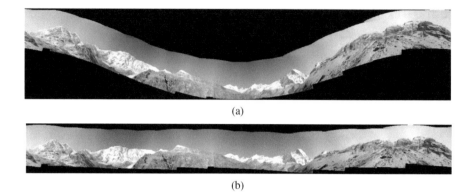

(a)

(b)

Fig. 2.1 Panorama straightening introduced by Brown and Lowe [11]. The figure shows (**a**) automatically generated panorama without straightening, and (**b**) the same panorama after the horizon straightening

image order, camera position, and scale. The algorithm also includes the automatic horizon detection that straightens the final image mosaic (shown in Fig. 2.1). Another method for image alignment is to use a video stream of frames [62].

Whereas image alignment determines the geometry of the image, the blending step handles the pixel intensity differences in the final mosaic. A major issue in creating photomosaics resides in the fact that the original images do not have identical brightness levels. This is usually caused by diverging camera orientations in space. Thus, some cameras acquire more light than the others. The problem manifests itself by the appearance of a visible seam in regions where the images overlap. The current algorithms can be divided into two categories: seam smoothing and optimal seam finding.

The seam smoothing algorithms reduce the color difference between two overlapped images to make the seamless noticeable and remove artifacts from the stitching process. The alpha blending [62, 100], i.e., blending based on a weighted average between pixels in every image, can reduce or even completely remove the seams. However, high frequency blurring may occur in the presence of any small image alignment error. If the alignment error occurs in the region where the objects can be found, it incurs the blending of a background pixel from one image with the foreground pixel from another. More advanced seam smoothing algorithms relate to the gradient domain image blending [22, 34, 43, 65, 86]. These algorithms use gradient domain operations to produce smoother color transitions and reduce color differences. Another possible solution to resolving color transition issue consists of using a multi-resolution blending algorithm [11, 14, 84] where high frequencies are combined in a small spatial range, thus avoiding blurring. Recently, a new approach has been proposed [69], where blending is performed using optical flow fields, providing high-quality mosaics.

Optimal seam finding approaches [1, 20, 101] search for seams in overlapping areas along paths where differences between source images are minimal. The seams can be used to label each output image pixel with the input image that should contribute to it. The combination of optimal seam finding and seam smoothing for image stitching has also been used in panorama applications [1]. Source images are first combined by compositing along optimal seams, and if the seams and stitching artifacts are visible, seam smoothing is applied to reduce color differences and hide the visible artifacts.

2.2   Camera Systems for Panorama Creation

2.2.1   Single Camera Systems

Early systems for capturing multiple views were based on a translating or rotating high-resolution camera for capturing and later rendering the scene [44, 56, 74, 79]. They are sometimes called *slit cameras*, since they use only short strips of the scene to generate a panoramic image. A computer-controlled motorized mechanism rotates the camera holder in order to acquire the full 360° view. The advantage of the rotating camera is in its capability to acquire a high-resolution omnidirectional image, however, at the cost of a long acquisition time. Therefore, it is difficult to use such systems to acquire a dynamic scene or a high frame rate video. Another disadvantage of these concepts is the limited vertical field of view, due to rotation around a single center. These problems were later resolved using a Pan-Tilt-Zoom (PTZ) camera [80], which is able to increase the vertical FOV by rotating around two axes (Fig. 2.2).

Recently, the researchers started including special applications to the single camera systems. Belbachir et al. [7, 8] designed a BiCa360 smart camera that reconstructs a sparse panorama for machine vision applications.

2.2.2   Catadioptric Systems

An alternate approach to omnidirectional acquisition is a catadioptric system [6], which consists of a convex mirror placed above a single camera sensor [55, 57]. The curved mirror of a catadioptric system is positioned in a way to reflect a 360° FOV environment directly into the lens. The mirror shape and lens are specifically chosen to allow a single viewpoint panorama, and easy unrolling of the acquired image to a cylindrical or spherical panorama.

Catadioptric systems have the advantage of real-time and high frame rate video acquisition. Furthermore, the mirror is used to divert the light into the lens. Hence, there is almost no chromatic aberrations or distortions, which are seen in wide FOV

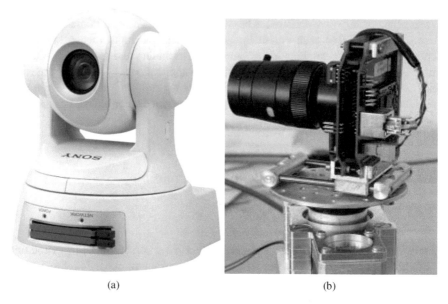

(a)   (b)

Fig. 2.2 Two examples of single camera systems placed on a rotational stand. (**a**) Commercial PTZ camera Sony SNC-RZ30, and (**b**) BiCa360 camera [8]

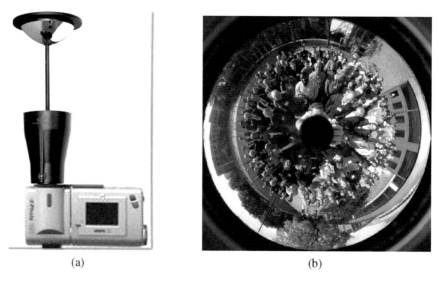

(a)   (b)

Fig. 2.3 An example of catadioptric systems (**a**) with lens and a curved mirror, and (**b**) its image output

images acquired using a *fish-eye* lens. However, catadioptric systems are limited to the resolution of the sensor. Furthermore, their overall FOV is limited, since it is restricted to the area below the sensor (Fig. 2.3).

2.2.3   Polydioptric Systems

A polydioptric (*poly*—"in multiple ways," *dioptric*—"vision assistance by focusing light") camera is a camera that captures a multi-perspective subset of the light ray space. In the scope of this book, discussion on polydioptric cameras will be restricted to multi-camera systems. The idea of using more than one camera has been first proposed by Taylor [90]. The use of linear array of still cameras allowed panoramic acquisition of dynamic scenes, which create ghosting effect in the rotating camera systems (Fig. 2.4).

Construction of panoramic videos requires large datasets, and researchers focused on developing systems with arrays of video cameras. The first camera array systems were built only for recording and later offline processing on PCs [67]. Other such systems [49, 58, 87, 105] were built with real-time processing capability for low resolution and low frame rates. A general-purpose camera array system was built by Wilburn et al. [99] with limited local processing at the camera level. This system was developed for recording large amounts of data and its intensive offline processing, and not for real-time operations. A similar system for creating high-resolution spherical panoramas was developed by Cogal et al. [17]. It was

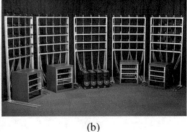

(b)

(a)

Fig. 2.4 (**a**) Polydioptric system designed at EPFL [17] that is able to acquire 220 Mpixels; (**b**) The Stanford multi-camera array [99]

envisioned for a high-resolution surveillance of both ground and aerial vehicles. Finally, two of the most well-known polydioptric systems for large data collection are the Google Street view [5] and the open-source design of Facebook Surround 360.

Recently, several systems were built for high-resolution acquisition with some sort of real-time processing. Schreer et al. [77] designed a multi-camera system for live video production, which includes multiple standard cameras, omnidirectional camera, and PTZ cameras. Xu et al. [102] built a system with six cameras placed on ring structure, providing high definition output in a limited FOV. The central direction of the FOV can be changed in real-time using sliding mechanism.

2.2.4   Commercial Cameras

Some of the modern digital cameras have the panoramic mode as a built-in function. They include a special capturing hardware and firmware, usually bracketed capture and gyroscope measurement, as well as embedded image stitching. Even some low-end cameras include this capability, e.g., Pentax Optio RZ10. Different names are given to this option: Motion Panorama (Fuji), Sweep Panorama (Sony), Easy Panorama (Nikon), or Panorama (Olympus). These are all single-lens cameras which take shots in a burst and stitch them into a single still panoramic image, i.e., they do not have a panoramic video capability.

With the advancement of technology, more and more companies started investing in integration of cameras and dedicated image processing chips to create what is now called a "Smart camera." As of today, all of the previously mentioned panorama generation techniques reached the market, and they are still available for purchase. A Swiss company Seitz produces several models of Roundshot cameras [78], which is a single-lens camera mounted on a rotating stand. Kogeto [37] sells a teleconference camera Jo, which is also based on a high-speed rotation of a single camera. Concerning multi-camera systems, the most popular ones are Pointgrey's Ladybug [66] (Ladybug5 is the latest one today) consisting of six CMOS image sensors, FullView [24] with a system including four cameras and flat mirrors, and Ricoh's THETA [70] that has two back-to-back sensors and fish-eye lenses.

FullView camera (Fig. 2.5b) is especially interesting since its four cameras are pointed at four flat mirrors looking outward. Effectively, each camera is looking in a different direction, but from the same single viewpoint. Reflection off a planar mirror is always clear and sharp, irrespective of the size of the camera aperture and its position relative to the mirror. As a result, composite images are artifact-free and blur-free. The drawback of this method is its reliability on mirrors and its limited field of view in the vertical direction.

Fig. 2.5 Commercially available panoramic cameras: (**a**) Pointgrey's Ladybug5, (**b**) Working principle of FullView, (**c**) Seitz Roundshot D3, and (**d**) Ricoh THETA

2.2.5   Light-Field and Unconventional Cameras

One of the latest types of computational cameras fall in the group of light-field cameras, i.e., they capture the complete light field in the environment. This gives a user an opportunity to render multiple views, unlike with the classical camera. There are several approaches to capture the light field.

Yang et al. [103] and Pelican Imaging cameras [63] capture the light field with a planar array of cameras. Commercially available cameras, such as Lytro [48] and Raytrix [68], use a microlens array and the main lens. The array is square-shaped and placed in the position where the standard cameras have the sensor. The main lens focuses the image on the microlens grid that is placed just in front of the sensor. The obtained image resembles the image of the microlens grid from the backside, and the true image is reconstructed using computational algorithms. This method allows post-processing operations such as refocusing, or the viewpoint change. The main drawbacks are lower resolution and low speed, due to the needed processing.

Recently, a camera system able to acquire an image frame with more than 1 gigapixel resolution was developed [10]. This camera uses a very complex lens system comprising of a parallel array of microcameras to acquire the image. Due to the extremely high resolution of the image, it suffers from a very low frame rate, even at low output resolution.

Some of the other interesting computational cameras include the work of Cossairt et al. [18] and Song et al. [83]. Cossairt et al. used multiple sensors arranged on a hemisphere looking inwards, and a single ball lens. This design also lacks the ability to process data with high frame rates. Song et al. designed a single camera in the shape of an arthropod eye, with the pixels manufactured on a stretchable surface. Due the manufacturing process and the technology limitations, this camera currently has only 256 pixels (Fig. 2.6).

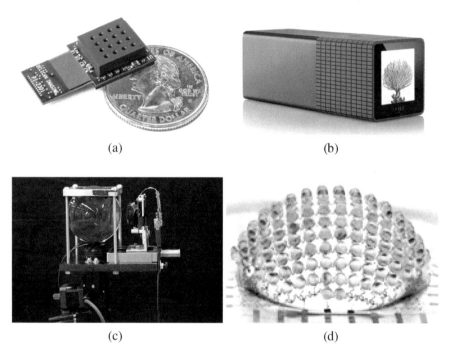

(a)   (b)

(c)   (d)

Fig. 2.6 Light-field cameras: (**a**) Pelican Imaging, (**b**) Lytro, (**c**) ball lens and a sensor array [18], and (**d**) arthropod-inspired camera [83]

2.3 Miniaturized Panoramic Camera Systems

2.3.1 Insect Eye-Mimicking Systems Based on Micro-Machining Techniques

This group of cameras is based on one-to-one morphological replica of the natural insect eyes. They usually try to mimic the natural counterparts by focusing on designing and replicating the smallest units, ommatidias, with the help of micro-machining methods and planar and/or curved electronic sensors.

The microlens array-based solutions can be classified into two main groups according to their shape as planar (2D) and curved (3D). The planar shape solutions [12, 13, 88, 89, 95] are limited in terms of field of view but they are easy to fabricate compared to the curved ones. Due to their planar nature, 2D systems cannot provide 360° FOV.

The other group in the literature, 3D or curved type microlens-based implementations, began to be developed more recently [23, 33, 41, 45, 83].

In [45], an optical solution with 3D microprism array and a 2D microlens array on top of a 2D sensor array is proposed. A complex and high precision micromachining processes such as regular diamond turning, diamond broaching, slow tool servo, and micromilling processes are utilized. The diameter of the optical system that was implemented was 20 mm. The authors did not report an assembly with a sensor array, but they made measurements with a charge coupled device (CCD) array-based camera with regular zoom optics.

A multi-aperture insect eye inspired system is described in [47]. The design is inspired by a housefly with a main objective of improving motion detection capability of insect eye inspired imaging systems. The design utilizes a lens-fiber optic pair to mimic each facet of the compound eye of the insects. Photodarlington devices are used as sensing elements that generate output current proportional to the gathered light. Although the authors reported improvement in motion detection capability, they reported a 7-pixel resolution in a diameter around 13 mm.

The work mentioned in the previous section [83] introduces a methodology for multi-lens curved imagers. The authors proposed integration techniques of elastomeric compound optical elements with thin silicon photo-detectors and a method for deforming these integrated sheets to get hemispherical surfaces. They reported a number of 180 bio-mimicked ommatidias, capable of imaging in a 360° × 80° field of view and physical dimensions around 12 mm in diameter. The authors compare their image capture capability with the eyes of fire ants (*Solenopsis fugax*) and bark beetles. They have also reported ray tracing simulation results but not real captured images with their proposed imaging system.

Another micro-machining approach is proposed in [23]. In this system, a 180° × 60° FOV is achieved with 630 ommatidias on a partial cylindrical surface. The authors reported inter-ommatidial angle of $\Delta\phi = 4.7°$ and acceptance angle of $\Delta\rho = 4.7°$. Their methods are based on complex integration and fabrication techniques. While the final spatial resolution is not notably large, the main advantage

Table 2.1 Summary of curved optics/electronics-based solutions for multi-aperture insect eye imaging

Solution	# of pixels	Field of view	Size (diameter) (mm)	Frame rate
[45]	N/A	$180° \times 180°$	20	N/A
[83]	180	$330° \times 330°$	12	N/A
[23]	630	$180° \times 60°$	12	300 fps

of this solution is reported as the high frame rate which allows optical flow estimation. The details of the system are also described in [96].

Both [83] and [23] are combination of micro-lens and micro-integration-based methods that are trying to copy natural insect eyes one to one basis. Therefore, as the insect eyes in nature have some limitations, the aforementioned methods also display similar diffraction limitation problem [38], which leads to low resolution with the given limited diameters [39, 60]. Some of these works are compared quantatively in their key aspect in Table 2.1.

2.3.2 Large FOV Imaging for Medical Endoscopy

There are many studies that show the need of large field of view, or panoramic imaging capability for endoscopy [21, 31, 46, 64, 71, 73, 98]. Most of these systems are optical solutions with either mirror or wide angle lens. The main drawbacks of these solutions are limited resolution, optical distortions, and lack of multi-perspective view. In [27], new technologies to overcome the narrow field of view problem in colonoscopy procedures are reviewed. One group of these emerging technologies in the market does not have the capability of giving a compact view [28, 72]. Hence, it leads to an increase in the diagnostics time [19]. The other group uses a mirror or a lens-based solution, and suffers from limited resolution capabilities, non-uniform resolution, and optical distortions [26, 32, 92]. The latter group does not offer multi-point view capability to have 3D imaging capabilities either.

In order to have a uniform and high-resolution imaging capability in such applications, the proposal stated in this book is to exploit the insect eye imaging to the endoscopy domain. Thus, we focus on developing a set of methods for creating miniature multi-camera system, which also have 3D image reconstruction capability. The illumination capability is an obvious requirement for endoscopic applications. In addition to the limitations of the previous insect eye inspired systems, there is no illumination method proposed for the insect eye inspired imaging systems to correctly work in a dark environment. Therefore, this requirement is also taken into account during the research and development phases presented in this book.

2.4 Depth Estimation Camera Systems and Approaches

Depth estimation can be performed by exploiting six main techniques: time-of-flight (TOF) camera, LIDAR sensors, radars, structured infrared light projection, learning-based single camera algorithms, and stereo camera.

A TOF camera easily measures the distance between the object and camera using a sensor and projection of light pulse, circumventing the need of intricate digital image processing hardware [42]. However, it does not provide efficient results when the distance between the object and camera is large. Moreover, the resolution of TOF cameras is usually very low (200 × 200) [42] in comparison to the Full HD display standard (1920 × 1080). Furthermore, their commercial price is much higher than the CMOS or CCD cameras.

LIDAR sensors are similar to the TOF cameras, but they compute the depth image using laser scanning mechanism [94]. LIDAR sensors are also very expensive. Due to laser scanning hardware, LIDAR sensors are heavy and bulky devices. However, laser scanner-based systems can measure the distance of the very far objects. Similarly, radars utilize radio waves to measure the distance of the far and large objects [81].

Structured infrared light projection-based depth measurement systems such as Microsoft Kinect [51] and Structure Sensor of Occipital [59] provide high-quality results even for the textureless objects and in dark environment. However, they are not able to provide high-quality results under sunlight due to interference of the infrared spectrum of the sun. Both Kinect and Structure Sensor provide depth estimation results in VGA resolution (640 × 480). Their resolution is lower than the CMOS and CCD cameras since the pixel size of infrared sensors is large, and their depth measurement range is limited by the projection power of the infrared beamer. For example, Kinect is not able to measure the distance of the object if they are further than 7 m. In addition, multiple structured infrared light projection systems cannot be oriented to the same location due to interference of the multiple projections.

Single camera-based depth estimation algorithms require very large training sets and learning approach [75]. Indeed, learning and intelligence are significant features of human vision that provides depth perception even with a single eye. However, implementing a high-quality and real-time depth estimation system is very challenging using a single camera since these algorithms are computationally intensive and require very large datasets. Consequently, in order to compute the depth map for real-time video processing applications, the majority of research focus on extracting the disparity information using two or more synchronized images taken from different viewpoints, using CMOS or CCD cameras [76].

Disparity estimation (DE)-based depth estimation systems can measure the distance of the objects even if the objects are very close or far by mechanically adapting distance between the stereo cameras. Multiple DE systems can be used in same environment since they don't project any light. In addition to indoor environment, they can be also used outdoor, since sunlight does not make interference problem.

However, implementation of real-time and high-quality DE for high-resolution (HR) video is challenging due to its computational complexity.

Many disparity estimation algorithms have been developed with the goal to provide high-quality disparity results. These are ranked with respect to their performance in the evaluation of Middlebury benchmarks [76]. Although top performing algorithms provide impressive visual and quantitative results [36, 50, 97], their implementations in real-time HR are challenging due to their complex multi-step refinement processes or their global processing requirements that demand huge memory size and bandwidth. For example, the AD-Census algorithm [50], currently the top published performer, provides successful results that are very close to the ground truth. However, this algorithm consists of multiple disparity enhancement sub-algorithms, and implementing them into a mid-range FPGA is very challenging both in terms of hardware resource and memory limitations.

Various hardware architectures that are presented in literature provide real-time DE [15, 25, 29, 35, 40, 52, 91]. Some implemented hardware architectures only target CIF or VGA video [15, 35, 40, 52]. The hardware proposed in [15] claims real-time only for CIF video. It uses the Census transform [104] and currently provides the highest quality disparity results compared to real-time hardware implementations in ASICs and FPGAs. The hardware presented in [15] uses the low-complexity Mini-Census method to determine the matching cost, and aggregates the Hamming costs following the method in [50]. Due to the high complexity of cost aggregation, the hardware proposed in [15] requires high memory bandwidth and large hardware resource utilization, even for low resolution (LR) video.

Real-time DE of HR images offers some crucial advantages compared to low resolution DE. First, processing HR stereo images increases the disparity map resolution which improves the quality of the object definition. Better object definition is essentially important for a variety of high-quality video processing applications such as object detection and tracking. Second, DE of HR stereo images offers the capacity to define the disparity with sub-pixel efficiency compared to the DE for LR image. Therefore, the DE for HR provides more precise depth measurements than the DE for LR.

Despite the advantages of HR disparity estimation, the use of HR stereo images presents some challenges. Disparity estimation needs to be assigned pixel by pixel for high-quality disparity estimation. Pixel-wise operations cause a sharp increase in computational complexity when the DE targets HR stereo video. Moreover, DE for HR stereo images requires stereo matching checks with larger number of candidate pixels than the disparity estimation for LR images. The large number of candidates increases the difficulty to reach real-time performance for HR images. Furthermore, high-quality disparity estimation may require multiple reads of input images or intermediate results, which poses severe demands on off-chip and on-chip memory size and bandwidth especially for HR images.

The systems proposed in [25, 29, 91] claim to reach real-time for HR video. Still, their quality results in terms of the HR benchmarks given in [76] are not provided. Georgoulas and Andreadis [25] claims to reach 550 fps for 80 pixel disparity range at a 800×600 video resolution, but it requires high amount of hardware resources. In addition, Georgoulas and Andreadis [25] claims that their hardware implementation operates at 511 MHz using a Stratix IV FPGA, however, without providing detail information related to the architecture and design that enable this high performance. A simple edge-directed method presented in [91] reaches 50 fps at a 1280×1024 video resolution and 120 pixel disparity range, but does not provide satisfactory DE results due to a low-complexity architecture. In [29], a hierarchical structure with respect to image resolution is presented to reach 30 fps at a 1920×1080 video resolution and 256 pixel disparity range, but it does not provide high-quality DE for HR.

More than two cameras can be used to improve the depth map. As presented in [54], trinocular DE solves most of the occlusion problems present in a single-pair camera system since any occluded region in a matched stereo pair (center-left) is not occluded in the opposite matched pair (center-right). Moreover, the double-checking scheme of trinocular DE improves binocular DE results even for unoccluded regions and provides correct disparity results even if the object is located in the left or right edge of the center image.

A limited number of trinocular disparity estimation hardware implementations are presented in the literature [16, 53]. The hardware presented in [53] enables handling 52 fps on an Altera Cyclone-IV FPGA at a 640×480 video resolution. The hardware presented in [16] uses a triangular configuration of three cameras and enables handling 30 fps on a Xilinx Virtex-4 FPGA at a 320×240 video resolution for a 64 pixels disparity range.

Camera calibration and image rectification are important pre-processing parts of the DE. The stereo matching process compares the pixels in the left and right images and provides the disparity value corresponding to each pixel. If the cameras could be perfectly aligned parallel and the lenses were perfect, without any distortion, the matching pixels would be located in the same row of the right and left images. However, providing a perfect setup is virtually impossible. Lens distortion and camera misalignments should be modeled and removed by internal and external stereo camera calibration and image rectification processes [9].

Many real-time stereo-matching hardware implementations [2, 15, 25] prove their DE efficiency using already calibrated and rectified benchmarks of the Middlebury evaluation set [76], while some do not provide detailed information related to the rectification of the original input images [29]. In a system that processes the disparity estimation in real-time, image rectification should also be performed in real-time. The rectification hardware implementation presented in [82] solves the complex equations that model distortion, and consumes a significant amount of hardware resources.

A look-up-table-based approach is a straightforward solution that consumes a low amount of hardware resources in an FPGA or ASIC [30, 61, 93]. In [30, 61, 93], the mappings between original image pixel coordinates and rectified image pixel coordinates are pre-computed and then used as look-up-tables. Due to the significant amount of generated data, these tables are stored in an external memory such as a DDR or SRAM [30, 93]. Using external storage for the image rectification process may increase the cost of the disparity estimation hardware system or impose additional external memory bandwidth limitations on the system. In [61], the look-up-tables are encoded to consume 1.3 MB data for 1280×720 size stereo images with a low-complexity compression scheme. This amount of data requires at least 295 Block RAMs (BRAM) without considering pixel buffers, thus it can only be supported by the largest Virtex-5 FPGAs or other recent high-end FPGAs. A novel algorithm which compresses the rectification information to fit the look-up-table into the on-chip memory of a Virtex-5 FPGA is presented in [3, 4]. This chapter mainly focuses on high performance disparity estimation implementations using binocular and trinocular camera systems. The readers are referred [3, 4] to obtain more information about novel rectification hardware implementations.

2.5   Conclusion

Solutions with multiple cameras offer higher and more uniform resolution than the single camera. Multiple cameras looking out directly into the scene from different positions see the world with inherently different perspectives. Hence each camera sees each point in space in a different direction. The difference in directions depends on the depth of the point of observation. This often makes it impossible to merge images into a single coherent image without image blending. The image blending is a process that requires substantial image overlap and is highly prone to cause artifacts such as blurring or ghosting. Nevertheless, a multi-camera system with overlapping fields of view can be utilized for other target application at the same time or independently, e.g., extending the dynamic range, or object tracking.

Most developed camera-array systems are bulky and not easily portable platforms. Their control and operation depend on multi-computer setups. In addition to synchronization of the cameras, very large data rates present new challenges for the implementation of these systems. In the following section, a portable real-time multi-camera system will be presented in detail.

References

1. Agarwala A, Dontcheva M, Agrawala M, Drucker S, Colburn A, Curless B, Salesin D, Cohen M (2004) Interactive digital photomontage. ACM Trans Graph 23(3):294–302. doi:10.1145/1015706.1015718
2. Akin A, Baz I, Atakan B, Boybat I, Schmid A, Leblebici Y (2013) A hardware-oriented dynamically adaptive disparity estimation algorithm and its real-time hardware. In: Proceedings of the 23rd ACM international conference on great lakes symposium on VLSI, GLSVLSI '13. ACM, New York, pp 155–160. doi:10.1145/2483028.2483082. http://doi.acm.org/10.1145/2483028.2483082
3. Akin A, Baz I, Manuel L, Schmid A, Leblebici Y (2013) Compressed look-up-table based real-time rectification algorithm and its hardware. In: Proceedings of the IFIP/IEEE international conference on VLSI-SOC

4. Akin A, Gaemperle LM, Najibi H, Schmid A, Leblebici Y (2015) Enhanced compressed look-up-table based real-time rectification hardware. VLSI-SoC: at the crossroads of emerging trends. Springer, Berlin

5. Anguelov D, Dulong C, Filip D, Frueh C, Lafon S, Lyon R, Ogale A, Vincent L, Weaver J (2010) Google street view: capturing the world at street level. Computer 43(6):32–38. doi:10.1109/MC.2010.170

6. Baker S, Nayar S (1998) A theory of catadioptric image formation. In: IEEE international conference on computer vision (ICCV), pp 35–42

7. Belbachir A, Pflugfelder R, Gmeiner R (2010) A neuromorphic smart camera for real-time 360° distortion-free panoramas. In: Proceedings of IEEE international conference on distributed smart cameras, pp 221–226

8. Belbachir A, Mayerhofer M, Matolin D, Colineau J (2012) Real-time 360° panoramic views using BiCa360, the fast rotating dynamic vision sensor to up to 10 rotations per sec. In: Proceedings of IEEE international conference on circuits and systems, pp 727–730. doi:10.1109/ISCAS.2012.6272139

9. Bouguet JY (2004) Camera calibration toolbox for matlab. http://www.vision.caltech.edu/bouguetj/ [Online]. Available: http://www.vision.caltech.edu/bouguetj/

10. Brady DJ, Gehm ME, Stack RA, Marks DL, Kittle DS, Golish DR, Vera EM, Feller SD (2012) Multiscale gigapixel photography. Nature 486(7403):386–389

11. Brown M, Lowe D (2007) Automatic panoramic image stitching using invariant features. Int J Comput Vis 74(1):59–73

12. Brückner A, Duparré J, Dannberg P, Bräuer A, Tünnermann A (2007) Artificial neural superposition eye. Opt Express 15(19):11922–11933. doi:10.1364/OE.15.011922. http://www.opticsexpress.org/abstract.cfm?URI=oe-15-19-11922

13. Brückner A, Duparré J, Leitel R, Dannberg P, Bräuer A, Tünnermann A (2010) Thin wafer-level camera lenses inspired by insect compound eyes. Opt Express 18(24):24379–24394. doi:10.1364/OE.18.024379. http://www.opticsexpress.org/abstract.cfm?URI=oe-18-24-24379

14. Burt P, Adelson E (1983) A multiresolution spline with application to image mosaics. ACM Trans Graph 2(4):217–236. doi:10.1145/245.247

15. Chang NC, Tsai TH, Hsu BH, Chen YC, Chang TS (2010) Algorithm and architecture of disparity estimation with mini-census adaptive support weight. IEEE Trans Circuits Syst Video Technol 20(6):792–805

16. Chen L, Jia Y, Li M (2012) An FPGA-based RGBD imager. Mach Vis Appl 23(3):513–525

17. Cogal O, Akin A, Seyid K, Popovic V, Schmid A, Leblebici Y (2014) A new omni-directional multi-camera system for high resolution surveillance. In: Proceeding of SPIE defense and security symposium, Baltimore, MD. doi:10.1117/12.2049698

18. Cossairt OS, Miau D, Nayar SK (2011) Gigapixel computational imaging. In: Proceedings of IEEE international conference on computational photography, pp 1–8

19. Dik VK (2015) Prevention of colorectal cancer development and mortality: from epidemiology to endoscopy. PhD thesis

20. Efros AA, Freeman WT (2001) Image quilting for texture synthesis and transfer. In: ACM SIGGRAPH 2001, pp 341–346

21. Elahi SF, Wang TD (2011) Future and advances in endoscopy. J Biophotonics 4(7–8):471–481

22. Farbman Z, Hoffer G, Lipman Y, Cohen-Or D, Lischinski D (2009) Coordinates for instant image cloning. ACM Trans Graph 28(3):1–9

23. Floreano D (2013) Miniature curved artificial compound eyes. Proc Natl Acad Sci 110(23):9267–9272

24. Fullview (2015) http://www.fullview.com. Accessed on 24 Oct 2015

25. Georgoulas C, Andreadis I (2009) A real-time occlusion aware hardware structure for disparity map computation. In: Image analysis and processing–ICIAP 2009. Springer, Berlin, pp 721–730

26. Gluck N, Fishman S, Melhem A, Goldfarb S, Halpern Z, Santo E (2014) Su1221 aer-o-scope™, a self-propelled pneumatic colonoscope, is superior to conventional colonoscopy in polyp detection. Gastroenterology 146(5, Suppl 1):S-406. http://dx. doi.org/10.1016/S0016-5085(14)61467-0, http://www.sciencedirect.com/science/article/pii/ S0016508514614670. 2014 {DDW} Abstract

27. Gralnek IM (2015) Emerging technological advancements in colonoscopy: Third Eye® Retroscope® and Third Eye® Panoramictm, Fuse® Full Spectrum Endoscopy® colonoscopy platform, Extra-wide-Angle-View colonoscope, and naviaidtm g-eyetm balloon colonoscope. Dig Endosc 27(2):223–231. doi:10.1111/den.12382. http://dx.doi.org/10.1111/den.12382

28. Gralnek IM, Carr-Locke DL, Segol O, Halpern Z, Siersema PD, Sloyer A, Fenster J, Lewis BS, Santo E, Suissa A, Segev M (2013) Comparison of standard forward-viewing mode versus ultrawide-viewing mode of a novel colonoscopy platform: a prospective, multicenter study in the detection of simulated polyps in an in vitro colon model (with video). Gastrointest Endosc 77(3):472–479. http://dx.doi.org/10.1016/j.gie.2012.12.011, http://www. sciencedirect.com/science/article/pii/S0016510712030647

29. Greisen P, Heinzle S, Gross M, Burg AP (2011) An FPGA-based processing pipeline for high-definition stereo video. EURASIP J Image Video Process 2011(1):1–13

30. Gribbon K, Johnston C, Bailey D (2003) A real-time FPGA implementation of a barrel distortion correction algorithm with bilinear interpolation. In: Image and vision computing, Palmerston North, pp 408–413

31. Gu Y, Xie X, Li G, Sun T, Zhang Q, Wang Z, Wang Z (2010) A new system design of the multi-view micro-ball endoscopy system. In: 2010 annual international conference of the IEEE Engineering in Medicine and Biology Society (EMBC). IEEE, New York, pp 6409–6412

32. Hasan N, Gross SA, Gralnek IM, Pochapin M, Kiesslich R, Halpern Z (2014) A novel balloon colonoscope detects significantly more simulated polyps than a standard colonoscope in a colon model. Gastrointest Endosc 80(6):1135–1140. http://dx.doi.org/10.1016/j.gie.2014.04. 024, http://www.sciencedirect.com/science/article/pii/S0016510714013923

33. Jeong KH, Lee LP (2006) Biologically inspired artificial compound eyes. Science 312(5773):557–561

34. Jia J, Sun J, Tang CK, Shum HY (2006) Drag-and-drop pasting. In: ACM SIGGRAPH, pp 631–637

35. Jin S, Cho J, Dai Pham X, Lee KM, Park SK, Kim M, Jeon JW (2010) FPGA design and implementation of a real-time stereo vision system. IEEE Trans Circuits Syst Video Technol 20(1):15–26

36. Klaus A, Sormann M, Karner K (2006) Segment-based stereo matching using belief propagation and a self-adapting dissimilarity measure. In: 18th international conference on pattern recognition, 2006. ICPR 2006, vol 3. IEEE, New York, pp 15–18

37. Kogeto J (2015) http://kogeto.com/jo.html. Accessed on 24 Oct 2015

38. Land MF (1997) Visual acuity in insects. Annu Rev Entomol 42(1):147–177

39. Land MF, Nilsson DE (2012) Animal eyes. Oxford University Press, Oxford

40. Lee SH, Sharma S (2011) Real-time disparity estimation algorithm for stereo camera systems. IEEE Trans Consum Electron 57(3):1018–1026

41. Lee LP, Szema R (2005) Inspirations from biological optics for advanced photonic systems. Science 310(5751):1148–1150

42. Lee C, Song H, Choi B, Ho YS (2011) 3d scene capturing using stereoscopic cameras and a time-of-flight camera. IEEE Trans Consum Electron 57(3):1370–1376. doi:10.1109/TCE.2011.6018896

43. Levin A, Zomet A, Peleg S, Weiss Y (2004) Seamless image stitching in the gradient domain. In: Computer vision - ECCV 2004. Lecture notes in computer science, vol 3024. Springer, Berlin, Heidelberg, pp 377–389. doi:10.1007/978-3-540-24673-2_31

44. Levoy M, Hanrahan P (1996) Light field rendering. In: Proceedings of the 23rd annual conference on computer graphics and interactive techniques, New York, SIGGRAPH '96, pp 31–42. doi:http://doi.acm.org/10.1145/237170.237199

45. Li L, Yi AY (2010) Development of a 3d artificial compound eye. Opt Express 18(17):18125–18137
46. Liu J, Wang B, Hu W, sun P, Li J, Duan H, Si J (2015) Global and local panoramic views for gastroscopy: an assisted method of gastroscopic lesion surveillance. IEEE Trans Biomed Eng (99):1–1. doi:10.1109/TBME.2015.2424438
47. Luke GP, Wright CH, Barrett SF (2012) A multiaperture bioinspired sensor with hyperacuity. IEEE Sensors J 12(2):308–314
48. Lytro (2015) http://lytro.com. Accessed on 24 Oct 2015
49. Majumder A, Seales W, Gopi M, Fuchs H (1999) Immersive teleconferencing: a new algorithm to generate seamless panoramic video imagery. In: Proceedings of the seventh ACM international conference on multimedia (Part 1), pp 169–178
50. Mei X, Sun X, Zhou M, Jiao S, Wang H, Zhang X (2011) On building an accurate stereo matching system on graphics hardware. In: 2011 IEEE international conference on computer vision workshops (ICCV workshops). IEEE, New York, pp 467–474
51. Microsoft (2008) Kinect. http://www.microsoft.com/en-us/kinectforwindows/ [Online]
52. Miyajima Y, Maruyama T (2003) A real-time stereo vision system with FPGA. In: Cheung PYK, Constantinides G (eds) Field programmable logic and application. Lecture notes in computer science, vol 2778. Springer, Berlin, Heidelberg, pp 448–457. doi:10.1007/978-3-540-45234-8_44. http://dx.doi.org/10.1007/978-3-540-45234-8_44
53. Motten A, Claesen L, Pan Y (2012) Trinocular disparity processor using a hierarchic classification structure. In: 2012 IEEE/IFIP 20th international conference on VLSI and system-on-chip (VLSI-SoC). IEEE, New York, pp 247–250
54. Mozerov M, Gonzàlez J, Roca X, Villanueva JJ (2009) Trinocular stereo matching with composite disparity space image. In: 2009 16th IEEE international conference on image processing (ICIP). IEEE, New York, pp 2089–2092
55. Nayar S (1997) Catadioptric omnidirectional camera. In: IEEE conference on computer vision and pattern recognition (CVPR), pp 482–488
56. Nayar S, Karmarkar A (2000) 360 x 360 Mosaics. In: IEEE conference on computer vision and pattern recognition (CVPR), vol 2, pp 388–395
57. Nayar SK, Peri V (1999) Folded catadioptric cameras. In: Proceedings of IEEE computer society conference on computer vision and pattern recognition, pp 217–223
58. Neumann U, Pintaric T, Rizzo A (2000) Immersive panoramic video. In: Proceedings of the eighth ACM international conference on multimedia, MULTIMEDIA '00. ACM, New York, pp 493–494. doi:10.1145/354384.376408. http://doi.acm.org/10.1145/354384.376408
59. Occipital (2012) Structure sensor. http://structure.io/ [Online]
60. Palka J (2006) Diffraction and visual acuity of insects. Science 149(3683):551–553
61. Park DH, Ko HS, Kim JG, Cho JD (2011) Real time rectification using differentially encoded lookup table. In: Proceedings of the 5th international conference on ubiquitous information management and communication. ACM, New York, p 47
62. Peleg S, Herman J (1997) Panoramic mosaics by manifold projection. In: IEEE conference on computer vision and pattern recognition, San Juan, Puerto Rico, pp 338–343. doi:10.1109/CVPR.1997.609346
63. Pelican Imaging (2015) http://pelicanimaging.com. Accessed on 24 Oct 2015
64. Peng CH, Cheng CH (2014) A panoramic endoscope design and implementation for minimally invasive surgery. In: 2014 IEEE international symposium on circuits and systems (ISCAS), pp 453–456. doi:10.1109/ISCAS.2014.6865168
65. Perez P, Gangnet M, Blake A (2003) Poisson image editing. ACM Trans Graph 22(3):313–318. doi:10.1145/882262.882269
66. Pointgrey (2015) Ladybug. https://www.ptgrey.com/360-degree-spherical-camera-systems. Accessed on 24 Oct 2015
67. Rander P, Narayanan PJ, Kanade T (1997) Virtualized reality: constructing time-varying virtual worlds from real world events. In: Proceedings of IEEE visualization '97, pp 277–284
68. Raytrix (2015) http://raytrix.de. Accessed on 24 Oct 2015

69. Richardt C, Pritch Y, Zimmer H, Sorkine-Hornung A (2013) Megastereo: constructing high-resolution stereo panoramas. In: Proceedings of IEEE conference on computer vision and pattern recognition (CVPR)
70. Ricoh (2015) THETA. https://theta360.com/. Accessed on 24 Oct 2015
71. Roulet P, Konen P, Villegas M, Thibault S, Garneau PY (2010) 360 endoscopy using panomorph lens technology. In: BiOS. International Society for Optics and Photonics, p 75580T
72. Rubin M, Bose KP, Kim SH (2014) Mo1517 successful deployment and use of third eye panoramic™ a novel side viewing video {CAP} fitted on a standard colonoscope. Gastrointest Endosc 79(5, Suppl):AB466. doi:http://dx.doi.org/10.1016/j.gie.2014.02.694. http://www.sciencedirect.com/science/article/pii/S0016510714008645. {DDW} 2014ASGE Program and Abstracts {DDW} 2014ASGE Program and Abstracts
73. Sagawa R, Sakai T, Echigo T, Yagi K, Shiba M, Higuchi K, Arakawa T, Yagi Y (2008) Omnidirectional vision attachment for medical endoscopes. In: The 8th workshop on omnidirectional vision, camera networks and non-classical cameras-OMNIVIS
74. Sarachik KB (1989) Characterising an indoor environment with a mobile robot and uncalibrated stereo. In: Proceedings of IEEE international conference on robotics and automation, pp 984–989. doi:10.1109/ROBOT.1989.100109
75. Saxena A, Chung SH, Ng AY (2008) 3-d depth reconstruction from a single still image. Int J Comput Vis 76(1):53–69
76. Scharstein D, Szeliski R (2002) A taxonomy and evaluation of dense two-frame stereo correspondence algorithms. Int J Comput Vis 47(1–3):7–42
77. Schreer O, Feldmann I, Weissig C, Kauff P, Schafer R (2013) Ultrahigh-resolution panoramic imaging for format-agnostic video production. Proc IEEE 101(1):99–114. doi:10.1109/JPROC.2012.2193850
78. Seitz (2015) Roundshot. http://www.roundshot.com. Accessed on 24 Oct 2015
79. Shum HY, He LW (1999) Rendering with concentric mosaics. In: Proceedings of the 26th annual conference on computer graphics and interactive techniques, SIGGRAPH '99. ACM Press/Addison-Wesley, New York, pp 299–306. doi:10.1145/311535.311573
80. Sinha SN (2014) Pan-Tilt-Zoom (PTZ) camera. In: Ikeuchi K (ed) Computer vision. Springer US, New York, pp 581–586. doi:10.1007/978-0-387-31439-6_496
81. Skolnik MI (1962) Introduction to radar. Radar handbook. McGraw-Hill, New York, p 2
82. Son HS, Bae Kr, Ok SH, Lee YH, Moon B (2012) A rectification hardware architecture for an adaptive multiple-baseline stereo vision system. In: Communication and networking. Springer, New York, pp 147–155
83. Song YM, Xie Y, Malyarchuk V, Xiao J, Jung I, Choi KJ, Liu Z, Park H, Lu C, Kim RH, Li R, Crozier KB, Huang Y, Rogers JA (2013) Digital cameras with designs inspired by the arthropod eye. Nature 497(7447):95–99. doi:10.1038/nature12083
84. Su MS, Hwang WL, Cheng KY (2001) Variational calculus approach to multiresolution image mosaic. In: Proceedings of international conference on image processing, vol 2, pp 245 –248. doi:10.1109/ICIP.2001.958470
85. Szeliski R, Shum HY (1997) Creating full view panoramic image mosaics and environment maps. In: Proceedings of the conference on computer graphics and interactive techniques. SIGGRAPH '97. ACM, New York, pp 251–258. http://dx.doi.org/10.1145/258734.258861
86. Szeliski R, Uyttendaele M, Steedly D (2011) Fast Poisson blending using multi-splines. In: IEEE international conference on computational photography (ICCP). doi:10.1109/ICCPHOT.2011.5753119
87. Tang WK, Wong TT, Heng PA (2005) A system for real-time panorama generation and display in tele-immersive applications. IEEE Trans Multimedia 7(2):280–292
88. Tanida J (2001) Thin observation module by bound optics (tombo): concept and experimental verification. Appl Opt 40(11):1806–1813

89. Tanida J, Kagawa K, Fujii K, Horisaki R (2009) A computational compound imaging system based on irregular array optics. In: Frontiers in optics 2009/laser science XXV/Fall 2009 OSA optics & photonics technical digest. Optical Society of America, p CWB1. doi:10.1364/COSI.2009.CWB1. http://www.osapublishing.org/abstract.cfm?URI=COSI-2009-CWB1

90. Taylor D (1996) Virtual camera movement: the way of the future? Am Cinematographer 77(8):93–100

91. Ttofis C, Hadjitheophanous S, Georghiades A, Theocharides T (2013) Edge-directed hardware architecture for real-time disparity map computation. IEEE Trans Comput 62(4):690–704

92. Uraoka T, Tanaka S, Matsumoto T, Matsuda T, Oka S, Moriyama T, Higashi R, Saito Y (2013) A novel extra-wide-angle—view colonoscope: a simulated pilot study using anatomic colorectal models. Gastrointest Endosc 77(3):480–483. http://dx.doi.org/10.1016/j.gie.2012.08.037, http://www.sciencedirect.com/science/article/pii/S0016510712026582

93. Vancea C, Nedevschi S (2007) Lut-based image rectification module implemented in FPGA. In: 2007 IEEE international conference on intelligent computer communication and processing. IEEE, New York, pp 147–154

94. Velodyne (2008) Hdl-g4e. http://velodynelidar.com/lidar/hdlproducts/hdl64e.aspx [Online]

95. Venkataraman K, Lelescu D, Duparré J, McMahon A, Molina G, Chatterjee P, Mullis R, Nayar S (2013) Picam: an ultra-thin high performance monolithic camera array. ACM Trans Graph 32(6):166:1–166:13. doi:10.1145/2508363.2508390. http://doi.acm.org/10.1145/2508363.2508390

96. Viollet S, Godiot S, Leitel R, Buss W, Breugnon P, Menouni M, Juston R, Expert F, Colonnier F, L'Eplattenier G et al (2014) Hardware architecture and cutting-edge assembly process of a tiny curved compound eye. Sensors 14(11):21702–21721

97. Wang ZF, Zheng ZG (2008) A region based stereo matching algorithm using cooperative optimization. In: IEEE conference on computer vision and pattern recognition, 2008. CVPR 2008. IEEE, New York, pp 1–8

98. Wang RCC, Deen MJ, Armstrong D, Fang Q (2011) Development of a catadioptric endoscope objective with forward and side views. J Biomed Opt 16(6):066015–066015

99. Wilburn B, Joshi N, Vaish V, Talvala EV, Antunez E, Barth A, Adams A, Horowitz M, Levoy M (2005) High performance imaging using large camera arrays. ACM Trans Graph 24:765–776. doi:10.1145/1073204.1073259

100. Xiong Y, Pulli K (2009) Mask-based image blending and its applications on mobile devices. In: SPIE multispectral image processing and pattern recognition (MIPPR), vol 7498. doi:10.1117/12.832379

101. Xiong Y, Pulli K (2010) Fast panorama stitching for high-quality panoramic images on mobile phones. IEEE Trans Consum Electron 56(2):298–306

102. Xu Y, Zhou Q, Gong L, Zhu M, Ding X, Teng R (2014) High-speed simultaneous image distortion correction transformations for a multicamera cylindrical panorama real-time video system using FPGA. IEEE Trans Circuits Syst Video Technol 24(6):1061–1069. doi:10.1109/TCSVT.2013.2290576

103. Yang JC, Everett M, Buehler C, McMillan L (2002) A real-time distributed light field camera. In: Proceedings of the 13th eurographics workshop on rendering, pp 77–86

104. Zabih R, Woodfill J (1994) Non-parametric local transforms for computing visual correspondence. In: Computer vision—ECCV'94. Springer, Berlin, pp 151–158

105. Zhang C, Chen T (2004) A self-reconfigurable camera array. In: Eurographics symposium on rendering, pp 243–254

Chapter 3
Panorama Construction Algorithms

In this chapter, we introduce panorama construction algorithms. These algorithms are developed for software processing on a PC, and their real-time implementation is often not possible. Thus, we provide modifications to the already existing algorithms, and develop a new one that can be implemented in hardware, and in real-time. The presented work results in improved image quality and the visually pleasant panorama for the human eye.

3.1 Fundamentals of Image Formation

The image formation can be approximated by the pinhole camera model for many computer vision applications [11, 20]. The pinhole camera projects a 3D world scene into a 2D image plane. In order to perform this projection, three coordinate systems are considered, as depicted in Fig. 3.1. The coordinates of point $\mathbf{X} = [x\ y\ z]^T$ are expressed in the world coordinate system with its origin at the point $\mathbf{O}$. The same point can also be expressed in the camera coordinate system by coordinates $\mathbf{X_{cam}}$. Origin of the camera coordinate system coincides with the camera's focal point $\mathbf{O_c} = \mathbf{C}$. The relation between the world and camera coordinates is unique and consists of a single translation $\mathbf{t}$ and a single rotation R:

$$\mathbf{X_{cam}} = R(\mathbf{X} - \mathbf{t}) \tag{3.1}$$

The vector $\mathbf{t}$ denotes the distance between origins of the world and the camera coordinate systems, whereas the rotation matrix R expresses three Euler rotations in order to align the mentioned systems' axes. Parameters R and $\mathbf{t}$ are called extrinsic camera calibration parameters.

© Springer International Publishing AG 2017
V. Popovic et al., *Design and Implementation of Real-Time Multi-Sensor Vision Systems*, DOI 10.1007/978-3-319-59057-8_3

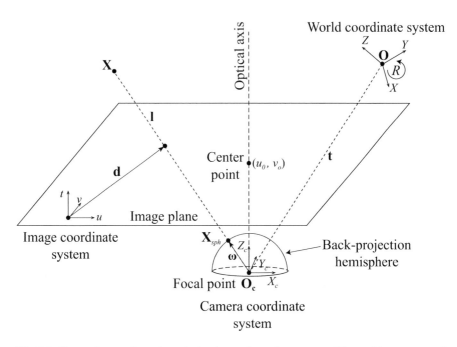

Fig. 3.1 Geometric transformations during image formation process. The world, camera, and image coordinate systems are shown with relations between them. A light ray passing through the world scene point **X** and the sensor's focal point intersects the image plane in point **d**, which represents a pixel in the acquired image. The back-projection procedure reconstructs the original light ray **l** and locates the intersection point with the back-projection hemisphere, $\mathbf{X}_{sph}$

The axes of the image coordinate system are aligned with the camera coordinate system. Hence, position of the projected point in the image coordinate system is expressed by:

$$\mathbf{d} = M(\mathbf{X} - \mathbf{c}) = K\mathbf{X}_{\mathbf{cam}} = \begin{bmatrix} f & 0 & u_0 \\ 0 & f & v_0 \\ 0 & 0 & 1 \end{bmatrix} \mathbf{X}_{\mathbf{cam}} \qquad (3.2)$$

where **d** is the pixel position in the image coordinate system, f is the focal length, u_0 and v_0 are camera center (principal) point offsets, M is the projection matrix, and K is the intrinsic calibration matrix. The origin of the system is coincident with one of the corners of the image sensor. Thus, the shift of the sensor's central point (u_0, v_0) to the sensor's corner is needed to obtain the final pixel position.

3.2 Image Stitching

Stitching of the panorama requires a correct alignment of all input images. In general, this can be done using uncalibrated system, e.g., mobile or handheld cameras, or using a stationary calibrated system. Most of the presented algorithm in Sect. 2.1 are automatic and work well in both system types. However, if the source image is noisy (low-light environment, low-quality sensor, etc.) or low resolution (specialized cameras, medical imaging), the automatic panorama generation often fails to correctly align and stitch the images. Figure 3.2 shows two examples of incorrect panorama generation using [4], due to insufficient number of feature points to make the proper matching. Both examples are panoramas generated from 15 source images taken by a stationary calibrated system, presented in [18].

This book focuses on fully calibrated systems and generating the panoramas as an omnidirectional view. For the sake of consistency, notation and the algorithms are explained for spherical view, i.e., projection of the source images onto the spherical surface, as shown in Fig. 3.1. The omnidirectional vision of a virtual observer located anywhere inside the hemisphere can be reconstructed by combining the information collected by each camera in the light ray space domain (or light field [13]).

In this process, the omnidirectional view is estimated on a discretized spherical surface S_d of directions. The surface of this hemisphere is discretized into a grid with N_θ latitude and N_ϕ longitude samples, where each sample represents one pixel. The hemisphere S_d corresponds to the back-projection hemisphere from Fig. 3.1. Figure 3.3a shows a discretized hemisphere with 16 pixels for N_θ and N_ϕ each. A unit vector $\omega \in S_d$, represented in the spherical coordinate system $\omega = (\theta_\omega, \phi_\omega)$, is assigned to the position of each pixel $\mathbf{X}_{\mathrm{sph}}$. Different pixel distributions over the hemisphere are discussed in Sect. 3.2.1.

The construction of the virtual omnidirectional view $\mathscr{L}(\mathbf{q}, \boldsymbol{\omega}) \in \mathbb{R}$, where $\mathbf{q}$ determines the location of the observer, consists of finding all light rays $\mathbf{l}$, and their respective projections onto S_d. This approach requires three processing

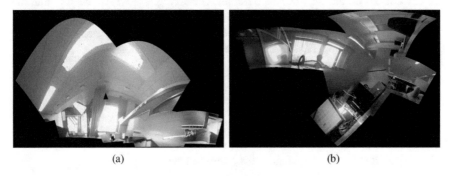

(a)   (b)

Fig. 3.2 Two examples of incorrect panorama generation from 15 input images, using the algorithm from [4]. The unexpected results are due to low number of feature points in the source images

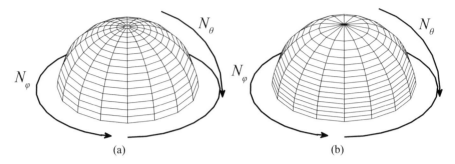

Fig. 3.3 Pixelized hemispherical surfaces S_d with $N_\theta = 16$ latitude pixels and $N_\phi = 16$ longitude pixels (total of 256 pixels) using (**a**) equiangular and (**b**) constant pixel density discretization

steps. The first step is to find all the cameras that capture the light ray **l**. The second step consists of finding a pixel in each input image that corresponds to the direction defined by ω. Finally, the third step consists of blending all pixel values corresponding to the same ω into one. The result is the reconstructed light ray intensity $\mathcal{L}(\mathbf{q}, \omega)$.

To reconstruct a pixel in the omnidirectional view, all images having the observed ω in their FOV are selected. Calibration of the system provides both intrinsic and extrinsic parameters [3]: focal length, center point offset, **t**, **u**, **v**, and angle-of-view α. The ω is within the camera's FOV if the following constraint is met:

$$\omega \cdot \mathbf{t_i} > \cos \frac{\alpha}{2}, \qquad i = 1 \ldots N_{cam} \qquad (3.3)$$

Figure 3.4a illustrates a hemispherical camera arrangement, where camera positions are marked with circles. The full circles represent the cameras that have the observed ω in their FOV. To extract the light intensity in that direction for each contributing camera, a pixel in the camera image frame has to be found. Due to the rectangular sampling grid of the cameras, the ω does not coincide with the exact pixel locations on the camera image frames. Observing Fig. 3.1, we need to find the closest pixel position **d** to the light ray **l**, i.e., ω. The exact position $\mathbf{d}' = [x, y]^T$ is obtained by finding the intersection point of **l** and the image frame. Using the pinhole camera model [11] and the assumption of a unit sphere:

$$\| \omega \|_2 = 1$$
$$\mathbf{d}' = (M \ \mathbf{l})^T \qquad (3.4)$$
$$\mathbf{d} = round(\mathbf{d}')$$

In Descartes world coordinates, this corresponds to:

$$(d'_x, d'_y) = -(\omega \cdot \mathbf{u}, \omega \cdot \mathbf{v}) \frac{f}{\omega \cdot \mathbf{t}} \qquad (3.5)$$

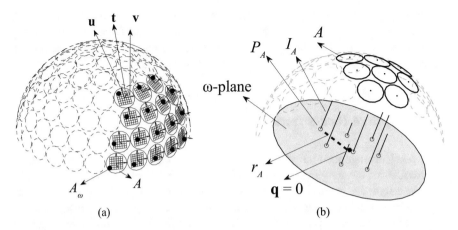

Fig. 3.4 (a) Cameras contributing to the direction $\boldsymbol{\omega}$ with their contributing pixels in the respective image frames, (b) projections of camera centers contributing in direction $\boldsymbol{\omega}$ onto planar surface perpendicular to $\boldsymbol{\omega}$. P_A represents the projected focal point of camera A and I_A represents the pixel intensity. r_A is the distance of the projection point of the camera center from the virtual observer $\mathbf{q}$

The pixel location is chosen using the nearest neighbor method, where the pixel closest to the desired direction is chosen as an estimate of the light ray intensity. The process is then repeated for all $\boldsymbol{\omega}$ and results in the pixel values $I(c_i, \boldsymbol{\omega})$, where c_i is the radial vector directing to the center position of the ith contributing camera. Figure 3.4a shows an example of the contributing cameras for an arbitrary pixel direction $\boldsymbol{\omega}$. The contributing position A_ω of the camera A, providing $I(c_A, \boldsymbol{\omega})$ is also indicated in Fig. 3.4a.

The third reconstruction step is performed in the space of light rays given by direction $\boldsymbol{\omega}$ and passing through the camera center positions. Under the assumption of Constant light flux (CLF), the light intensity remains constant on the trajectory of any light ray. Following the CLF assumption, the light ray intensity for a given direction $\boldsymbol{\omega}$ only varies in its respective orthographic plane. The orthographic plane is a plane normal to $\boldsymbol{\omega}$. Such plane is indicated as the "ω-plane" in Fig. 3.4b, and represented as a gray-shaded circle (the boundary of the circle is drawn for clarity purposes). The light ray in direction $\boldsymbol{\omega}$ recorded by each contributing camera intersects the ω-plane in points that are the projections of the cameras focal points on this plane. The projected focal points of the contributing cameras in $\boldsymbol{\omega}$ direction onto the ω-plane are highlighted by hollow points in Fig. 3.4b. Each projected camera point P_i on the planar surface is assigned the intensity value $I(c_i, \boldsymbol{\omega})$.

As an example, the projected focal point of camera A onto the ω-plane (i.e., P_A) in Fig. 3.4b is assigned the intensity value I_A. The virtual observer point inside the hemisphere (i.e., $\mathbf{q}$) is also projected onto the ω-plane. The light intensity value at the projected observer point (i.e., $\mathscr{L}(\mathbf{q}, \boldsymbol{\omega})$) is estimated by one of the blending algorithms, taking into account all $I(c_i, \boldsymbol{\omega})$ values or only a subset of them. In the given example, each of the seven contributing cameras shown with bold perimeter in Fig. 3.4b provides an intensity value which is observed in direction $\boldsymbol{\omega}$ for the observer position $\mathbf{q} = 0$. The observer is located in the center of the sphere and indicated by a bold dot. A single intensity value is calculated from the contributing intensities through a blending procedure on its respective ω-plane. The process is repeated for all $\boldsymbol{\omega}$ directions to create a full $360°$ panorama.

3.2.1 Sphere Discretization

The pixel arrangement $\boldsymbol{\omega}$ shown in Fig. 3.3a is derived from an equiangular distribution of the longitude and the latitude coordinates of a unit sphere into N_ϕ and N_θ segments, respectively. The equiangular distribution is defined by equal longitude and latitude angles between two neighboring pixels. This discretization enables the rectangular presentation of the reconstructed image suitable for standard displays, but results in the unequal contribution of the cameras. The density of the pixel directions close to the poles of the sphere is higher compared to the equator of the sphere in the equiangular scheme. Hence, the cameras positioned closer to the poles of the sphere contribute to more pixels in comparison to the other cameras in the system. The equiangular distribution is derived mathematically from (3.6):

$$\phi_\omega(i) = \frac{2\pi}{N_\phi} \times i, \quad 0 \le i < N_\phi$$

$$\theta_\omega(j) = \frac{\pi}{2N_\theta} \times \left(j + \frac{1}{2}\right), \quad 0 \le j < N_\theta$$

(3.6)

Spherical panoramas often have more details around the equator than on the poles. Furthermore, the systems with camera arrangement as in Fig. 3.4a place more cameras in the bottom rows, which allows higher spatial resolution. A constant pixel density scheme results in an approximately even contribution of the cameras. The scheme is based on enforcing a constant number of pixels per area, as expressed in (3.7). Compared to the equiangular distribution, the difference is observed in latitude angles. The discretization scheme expressed in (3.8) is derived by solving the integral in (3.7). The illustration of the constant pixel density sphere is shown in Fig. 3.3b.

$$\frac{N_\phi \times j}{\int_0^{2\pi} d\phi \int_0^{\theta_\omega(j)} \sin\theta \, d\theta} = \frac{N_\phi \times N_\theta}{2\pi}, \quad 0 \le j < N_\theta$$

(3.7)

$$\phi_\omega(i) = \frac{2\pi}{N_\phi} \times i, \quad 0 \le i < N_\phi$$

$$\theta_\omega(j) = \arccos\left(1 - \frac{j}{N_\theta}\right) + \theta_0, \quad 0 \le j < N_\theta.$$

(3.8)

The offset value θ_0 is added to the latitude angle in (3.8) to avoid repetition of pixel direction for the $j = 0$ case.

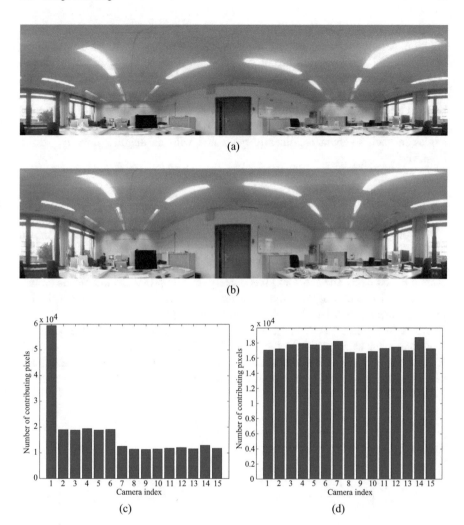

Fig. 3.5 A computer laboratory at the Swiss Federal Institute of Technology in Lausanne (EPFL, ELD227). Panoramic construction with a pixel resolution of $N_\phi \times N_\theta = 1024 \times 256$ (**a**) using the equiangular discretization, (**b**) using constant pixel density discretization. Figures (**c**) and (**d**) show the number of pixels each camera contributes to in case of (**c**) equiangular and (**d**) constant pixel density discretization

Figures 3.5a and 3.6b illustrate the differences between two discretization methods in panorama construction, and Figures 3.5c and 3.5d show the per-camera pixel contribution. Camera contribution in the constant pixel density distribution is almost equal for all cameras, whereas the top camera (looking in the north pole direction) in equiangular distribution contributes three times more than any other camera.

Latitudinal pixel distribution does not need to be a linear or a trigonometrical function. Moreover, it can be any function, including any of the piecewise linear (PWL) ones. PWL functions are of special interest, since the pixel emphasis can be placed on several places on the sphere. To achieve such discretization, the full latitudinal FOV of $\pi/2$ is divided into M pieces of arbitrary FOV_i, where each piece is linearly sampled. A number of pixels p_i are chosen for each of the pieces, separately, based on the desired application and view specifications. The latitude angles in each segment are linearly generated with an angular slope expressed in (3.9):

$$\Delta\theta_i = \frac{FOV_i}{p_i}, \quad 1 < i \leq M \tag{3.9}$$

A comparison between the presented pixel distribution schemes is shown in Fig. 3.6. An arbitrary PWL function comprising $M = 3$ pieces is taken for illustration purposes. This function results in higher pixel density near the pole and around the equator. The constant pixel density scheme provides more pixels around the equator, i.e., when latitude angles are higher. Finally, the equiangular distribution provides linearly distributed pixels around the hemisphere.

Apart from region selectivity, the PWL scheme is used for approximation of functions such as logarithms or exponentials. The PWL approximation of such functions is necessary for a compact hardware implementation. These functions can be used when more details are required around the pole or the equator, respectively.

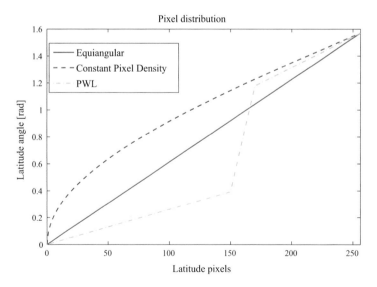

Fig. 3.6 Latitude angle distribution for $N_\theta = 256$ latitude pixels using three different pixel distribution schemes

3.2.2   Grid Refinement

The presented discretization schemes can be regarded as sampling grids of the surrounding light field. The total number of acquired pixels linearly increases with the number of cameras. Thus, the light field can become oversampled using only several low-resolution cameras. Light-field information is obtained at the sub-pixel scale as a benefit of this particular light-field oversampling. Hence, the acquired images contain finer detail than shown on a fixed-grid rendered panorama. In addition to the fact that the resolution of the reconstructed image can be significantly smaller than the total number of acquired pixels, this creates an excess of pixels that are not used in the reconstruction process.

Nevertheless, the acquisition of the excess pixels can be useful. If an ω direction in the reconstructed image is observed by more than one camera, i.e., parallax[1] exists in each point in space, a sub-pixel resolution can be achieved.

As presented in Sect. 3.2.1, it is possible to change the pixel distribution schemes. Additionally, the desired FOV is also programmable. Hence, a constant output resolution with the reduced FOV results in a grid refinement effect. The example of the refined pixelization is shown in Fig. 3.7, where the increase in pixel density can be noticed in the desired FOV.

The effect observed in the reconstructed image is similar to the effect of digital zoom. However, the sub-pixel data is taken from the real and previously unused measured data, i.e., it is not calculated using an interpolation function as in digital zooming. Hence, grid refinement provides a more truthful light field rendering than digital zoom, and it is shown in Fig. 3.8.

Fig. 3.7 Refined pixelization scheme with $N_\theta = 16$ latitude pixels and $N_\phi = 16$ longitude pixels. Longitudinal FOV is reduced to a quarter of the hemisphere

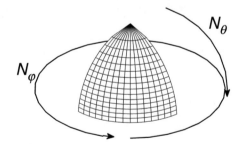

---

[1]Parallax is a displacement or difference in the apparent position of an object viewed along two different lines of sight.

Fig. 3.8 Detailed image parts obtained using Gaussian blending with the grid refinement: a lamp magnified 8×, the books magnified 32×, a desk magnified 8×

Fig. 3.9 The image on *left* shows the vignetting effect as dark regions around the edges and corners of the image. The image on the *right* is the same image after the correction is applied [Online source: http://intothenightphoto.blogspot.ch]

3.3 Vignetting Correction

Vignetting is an adverse effect observed in cameras, where the pixels located close to the image frame borders are significantly darker than the pixels located in the center. Vignetting also affects the reconstructed omnidirectional view; thus, pixel intensities in the reconstructed image alternatively vary, i.e., certain regions are darker and others are brighter. An example of vignetting effect is shown in Fig. 3.9, where the left and the right image represent images before and after correction, respectively.

Several methods are proposed in literature for modeling the vignetting effect and its correction. The chosen model for Panoptic camera is the Kang–Weiss model [12]. The Kang–Weiss model takes into account the pixel position in the camera image frame, the camera focal length, and a camera constant named the vignetting factor.

All pixels in each camera frame are corrected by multiplying the sampled pixel intensity with a correction factor. The corrected pixel intensity is expressed as:

$$I'(u, v) = I(u, v)(1 - \alpha d)\frac{1}{(1 + (d/f)^2)^2} \tag{3.10}$$

where α is the vignetting factor, f is the focal length, $I(u, v)$ is the original pixel intensity at coordinates (u, v) and $d = \sqrt{u^2 + v^2}$.

3.4 Alpha Blending

The first step in omnidirectional vision construction discussed in Sect. 3.2 consists of determining contributing pixels from input image frames and their respective intensities, $\mathscr{L}(c_i, \omega)$. Ideally, intensity values corresponding to the same direction will be equal in all the images. However, various parasitic effects, such as fabrication process, lens distortion, or exposure difference, influence difference in intensity. Furthermore, the stitching algorithm detailed in the previous section assumes that all the objects are far away from the camera. In reality, this is not always true, and the pixel intensities may differ significantly. Also, the obtained values may significantly vary due to diverging camera orientations, difference in light incidence angle, and misalignment of pixels due the calibration procedure. Even though the vignetting correction equalizes brightness of the individual camera's image, the reconstructed image quality mostly depends on the blending algorithm.

Alpha blending is a simple blending process where pixel intensities $\mathscr{L}(c_i, \omega)$ from all contributing cameras are weight-averaged. Different blending effects are obtained by changing the weights.

A special case of the alpha blending is the nearest neighbor (NN) blending. When applying the NN blending, the light intensity at the virtual observer point for each ω direction is set to the light intensity value of the best observing camera for that direction. In terms of weighted average, this corresponds to the case when the weight of only one camera is 1, while the other cameras contribute with a 0 weight. The NN blending is expressed in (3.11) in mathematical terms:

$$\mathscr{L}(\mathbf{q}, \omega) = \mathscr{L}(c_j, \omega)$$
$$j = \text{argmin}_{i \in I}(r_i) \tag{3.11}$$

where $I = \{i | \omega \cdot \mathbf{t_i} \geq \cos(\frac{\alpha_i}{2})\}$ is the index of the subset of contributing cameras for the pixel direction ω. A pixel direction ω is assumed observable by the camera c_i if the angle between its focal plane vector $\mathbf{t_i}$ (see Fig. 3.4a) and the pixel direction ω is smaller than half of the angle of view α_i of camera c_i. The length r_i identifies the distance between the projected focal point of camera c_i and the projected virtual observer point on the ω-plane. The camera with the smallest r distance to the projected point of virtual observer on the ω-plane is considered the best observing camera. As an illustration, such distance is identified with r_A and depicted by a dashed line for the contributing camera A in Fig. 3.4b.

The NN blending does not resolve intensity differences between pixels from different images, and results in clearly visible seams [16, 18, 19]. The resulting image examples can be seen in Fig. 3.10a. The artifacts caused by different brightness levels between cameras and misalignment can be resolved to a certain extent using linearly distributed weights.

The linear weighting incorporates all the cameras contributing into a selected ω direction through a linear combination [1, 19]. This is conducted by aggregating the weighted intensities of the contributing cameras. The weight of a contributing camera is the reciprocal of the distance between its projected focal point and the projected virtual observer point on the ω-plane, i.e., r_A in Fig. 3.4b. The weights are also normalized to the sum of the inverse of all the contributing camera's distances. It is mathematically expressed in (3.12):

$$\mathscr{L}(\mathbf{q}, \boldsymbol{\omega}) = \frac{\sum\limits_{i \in I} w_i \cdot \mathscr{L}(c_i, \boldsymbol{\omega})}{\sum\limits_{i \in I} w_i} \tag{3.12}$$

$$w_i = \frac{1}{r_i}$$

3.5 Gaussian Blending

The NN and linear weighting present several issues. An image reconstructed using the NN method shows clear boundaries between the fields of view of different cameras. Although some brightness differences are reduced by the vignetting correction, the boundaries are still visible and create an unpleasant effect to the human eye.

Linear weighting solves the problem of sharp boundaries to a certain extent. Pixels in the regions where cameras' fields of view overlap are blended using a weighted average, as expressed in (3.12). The intensity difference is reduced, but it is still existent. Moreover, the main disadvantage lies in the appearance of blurred edges in the image due to the misalignment and linearly chosen weights.

We propose a twofold modification of the alpha blending algorithm. The first modification relates to weights being a function of not only the camera's physical orientation $\mathbf{t}$, but also of the camera's intrinsic parameters and the position of the virtual observer. This modification is realized by adding a multiplicative factor to the linear alpha blending that is dependant on the pixel position d within the frame.

The second modification concerns the calculation of the multiplicative factor. By conducting subjective tests, it was empirically deduced that distributing the weights

(a)

(b)

(c)

(d)

Fig. 3.10 A computer laboratory at the Swiss Federal Institute of Technology in Lausanne (EPFL, ELD227). Panoramic construction with a pixel resolution of $N_\phi \times N_\theta = 1024 \times 256$ (**a**) using the nearest neighbor technique, (**b**) using alpha blending, (**c**) using Gaussian blending with $\sigma_d = 100$ and (**d**) using Adaptive Gaussian blending with $\sigma_d = 100$ and $\sigma_r = 1/30$

using a Gaussian function with respect to the pixel distance from the frame center results in the best image mosaic quality [18]. The image quality is improved in terms of both reducing visibility of the seams and reducing the ghosts around edges of the scene objects. The new weights in the weighted average expression are:

$$w_{i,j} = \frac{1}{r_i} \cdot \mathcal{G}(d_j, \sigma_d)$$

$$\mathcal{G}(d_j, \sigma_d) = e^{-\frac{d_j^2}{2\sigma_d^2}}$$

(3.13)

where r_i is the same distance as in (3.12), d_j is the distance of the jth pixel in the input image frame from its center, and σ_d is the variance of the Gaussian distribution function $\mathcal{G}$.

By adding the Gaussian factor to the weighted average expression, the seams between cameras' FOVs are not visible any more, as shown in Fig. 3.10c. Furthermore, the Gaussian blending reduces the difference in brightness in the images from different cameras and the overlapping regions are equalized with their respective surroundings. High-frequency blur is also reduced compared to the linear alpha blending weights.

3.5.1 Adaptive Gaussian Blending

The NN blending proves to be suitable for processing the pixels which are close to the camera center. Towards the boundaries of the camera's FOV, Gaussian blending is favorable, thanks to the brightness equalization and reduction of effects originating from the camera misalignment. The Adaptive Gaussian (AG) blending technique aims to restrict the Gaussian blending to the areas where the reconstructed pixels are not close to the center of a single camera's FOV. The NN blending is used in the areas close to the mentioned centers. Hence, this method benefits from the advantages of both Gaussian and NN blending.

One way of implementing this method consists of simultaneously constructing the two views and blending them for the output display. However, the implementation of such approach is computationally demanding, and its real-time implementation is not possible with the current technology.

An efficient implementation of the AG blending is proposed. A new confidence factor is introduced, which is related to each camera's observation of a given ω direction. For that purpose, a dot product of the ω and the focal vector $\mathbf{t}$ (see Fig. 3.4a) is taken as a reference metric.

In the blending phase of the reconstruction, a Gaussian confidence factor with respect to its $\omega \cdot \mathbf{t}_i$ is multiplied with the previously calculated $w_{i,j}$ of each camera c_i obtained from the Gaussian blending technique. By expanding (3.13), the AG blending weight and the Gaussian confidence factor are expressed in mathematical terms:

$$\tilde{w}_{i,j} = \frac{1}{r_i} \cdot \mathscr{G}(d_j, \sigma_d) \cdot \mathscr{C}(\boldsymbol{\omega}, c_i)$$

$$\mathscr{C}(\boldsymbol{\omega}, c_i) = e^{-\frac{(\boldsymbol{\omega} \cdot \mathbf{t}_i - 1)^2}{\sigma_r^2}}$$

(3.14)

where $\tilde{w}_{i,j}$ represents the new blending weight for jth pixel in the ith camera frame and $\mathscr{C}$ represents the AG confidence factor.

The AG blending favors very high values of $\boldsymbol{\omega} \cdot \mathbf{t}$ for a single camera. High values represent pixels which are positioned around the center of the camera frame. These pixels are considered to be more reliable than the ones located on the borders of the frame. In practice, the majority of $\boldsymbol{\omega}$ have one dominant camera, i.e., these pixels will be around the frame center of only one camera. Thus, the AG blending should neutralize the effects of all other cameras by assigning them a very low confidence factor and keeping only the dominant camera, similar to NN blending. In situations when an $\boldsymbol{\omega}$ has more than one high value of $\boldsymbol{\omega} \cdot \mathbf{t}$, the confidence factors adapt the blending weights to use more than one contributing camera in the weighted average, resembling the Gaussian blending.

The proposed AG blending implementation does not visually differ from the approach consisting of creating two views. Furthermore, the standard deviation of the confidence factor can be manually adapted to obtain the best possible image quality.

An example of the AG confidence factor curve is shown in Fig. 3.11, using σ_r set to $\frac{1}{30}$. The regions drawn over the curve depict the restrictions imposed on the Gaussian blending to obtain the AG blending. NN blending is applied in regions where the confidence factor is higher than 0.9, or almost 0, while Gaussian blending is applied in the transition region. This division reduces the influence of low-confidence over high-confidence pixels. Thus, the image mosaic is sharp in areas close to a single cameras' center, while the camera overlapping regions located on the periphery are blended using a Gaussian weight distribution.

3.6 Multi-Band Blending

Multi-band blending (MBB) [4] is based on a multi-resolution decomposition of the original images and their blending across octave frequency bands. The images are represented using a Laplacian pyramid (LP) [7], as it has perfect and simple reconstruction [22]. Several steps are performed to obtain the desired LP.

Laplacian pyramid is a multispectral and multiscale representation of an image, where each pyramid level contains one frequency band of the image. Image decomposition, processing, and reconstruction using a four-level LP are illustrated in Fig. 3.12. Let x be the source image. The image is filtered using the analysis low-pass filter $\mathbf{H}(\mathbf{z})$ and downsampled by two, in both horizontal and vertical directions. The decimated image is then upsampled and filtered with the synthesis filter $\mathbf{G}(\mathbf{z})$.

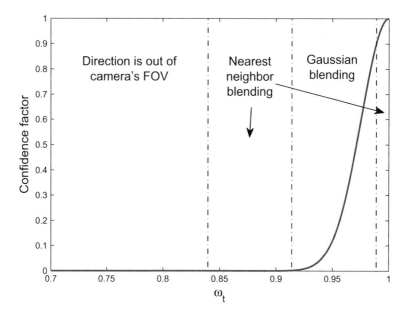

Fig. 3.11 Confidence factor based on $\omega_t = \omega \cdot t$. Gaussian blending is applied only in the region where the confidence factor is lower than 0.9

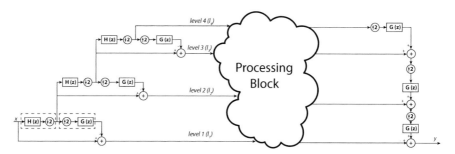

Fig. 3.12 Laplacian pyramid decomposition and reconstruction with four levels. The top level (*level 4*) represents the coarse approximation, whereas the bottom level represents the details. The analysis (**H(z)**) and synthesis (**G(z)**) filters with internal downsampling are marked with *red* and *blue dashed rectangles*. They are marked only in the first level for clarity reasons. The processing block denotes operations on the decomposed pyramid. The LP reconstruction is performed on the processed pixels, on the right side of the figure

The first level (l_1 in Fig. 3.12) of the LP is obtained by subtraction of the interpolated image from the source x, and it represents the high-frequency content of the image, i.e., the details. The decimated image is also used as the source for the second level of decomposition. An L-level LP is created using $L - 1$ repetitions of the mentioned principle.

The regions of overlap between images may be of an arbitrary shape. Thus, a mask should be created, defining the pixels which should be taken from the original image and their respective weight [7]. A binary mask is assigned to each image, where value 1 represents a pixel that should be taken from the selected image. This mask is further decomposed into a Gaussian pyramid (GP), which is created by repetitive blur and downsample operations, i.e., each level of the pyramid is a low-pass version of the previous level.

Each frequency band of the LP is combined with the respective frequency band of the other LPs, i.e., other images. A weighted average is applied within the overlapped areas, which are proportional in size to the wavelengths represented in the band. Hence, when coarse features occur in the overlapping region, they are gradually blended over a relatively large distance without blurring or degrading finer image details in the neighborhood [7]. The weights are taken from the corresponding mask GP. In case of blending two images, image A and image B, the blending of one pyramid level is expressed as:

$$I(x, y) = I_A(x, y)w(x, y) + I_B(x, y)(1 - w(x, y)) \tag{3.15}$$

where I_A and I_B are pixel intensities and w is the pixel weight. Figure 3.13 illustrates MBB procedure of N source images into a single panorama.

3.6.1 Choice of Filters

Quality of the blended panorama relates to the chosen set of filters for MBB. Only finite impulse response (FIR) filters are considered, since they are zeros-only filters and they can be efficiently implemented in hardware using convolution. A filter dataset of five low-pass filters is designed, and their performance is analyzed. The dataset consists of:

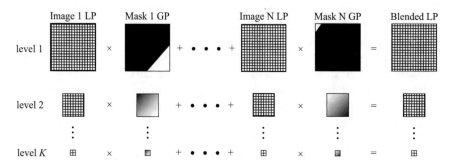

Fig. 3.13 An illustration of multi-band blending of N images using decomposition into a K-level Laplacian pyramid

- 5-tap binomial filter [22]
- 5/3 LeGall filter
- 9/7 Daubechies filter
- Custom 9-tap maximally flat filter
- Custom 9-tap equiripple filter.

The constraints on the custom-designed filters are driven by the hardware complexity and the frequency specifications of the first three filters. Hardware complexity and area utilization increase with the filter order. Additionally, high-order filters require large buffers for temporary pixel storage. Thus, we derive constraints which fit specifications of all compared filters. The designed filters are maximum eighth order, the cut-off frequency is $\pi/2$, and passband and stopband attenuations are 0.03 dB and 70 dB, respectively. The designed filter is used to create an orthogonal filter set from which decomposition and reconstruction filters are used as $\mathbf{H(z)}$ and $\mathbf{G(z)}$. These filters guarantee a perfect reconstruction of a single image, i.e., without blending. Thus, the loss of image quality in the reconstructed panorama pertains only to the stitching process and it is not a consequence of a faulty reconstruction.

Brown [4] and Burt [7] suggest to use the same filter for generation of the GP weight mask as for the LP. The use of the same filter simplifies the system and provides seamless blending results when the overall brightness level of the images does not differ significantly. However, when the images are taken with different exposure levels, the seam becomes noticeable [17].

To solve this issue, we propose the use of linearly distributed weights, instead of a Gaussian distribution. The proposed filter to create the GP of weights is:

$$H(z) = \frac{1}{5}(1 + z^{-1} + z^{-2} + z^{-3} + z^{-4}) \qquad (3.16)$$

The selected filter set for MBB was tested on a database of images, partly provided by A. Goshtasby from Image Fusion Systems Research [10]. Multiple shots of the same scene have been captured with different exposure times. More image sets were created by slightly shifting the image horizontally and diagonally, to emulate errors in registration that occur in real-life conditions. The benchmark database consisted of 30 different blended images and all of the input images had different exposure levels. One third had no registration errors, one third had horizontal shifts, and one third had diagonal shifts. Objective image quality metrics were applied to compare the quality of blending. A set of objective metrics based on both perceptual visual quality and statistical properties of the image is determined. The set consisted of the No-Reference Blur Metric (NRBM) [14], the Edge Quality (QE) [25], and the naturalness index (NIQE) [15]. For color images, the values were measured on the luminance component. The photomosaics in these cases were created using the maximum number of pyramidal levels of decomposition. Table 3.1 shows the obtained results for four different scenes. The best result for a scene is marked in bold.

Table 3.1 Objective quality metric comparison

Image	Filter	Metric		
		NRBM	QE	NIQE
Room	Binomial	5.23	0.6568	**8.2810**
	5/3	**4.37**	**0.7368**	8.2840
	9/7	**4.37**	0.7119	8.2944
	Maxflat	4.59	0.7288	8.2842
	Equiripple	4.52	0.7302	8.3066
House	Binomial	4.11	0.6910	3.1445
	5/3	**3.85**	**0.7422**	3.1390
	9/7	3.86	0.7400	3.1435
	Maxflat	3.88	0.7403	**3.1365**
	Equiripple	3.87	0.7399	3.1370
Mountains	Binomial	3.62	0.8024	2.7472
	5/3	**3.61**	0.8575	**2.7273**
	9/7	**3.61**	0.8560	2.7441
	Maxflat	3.62	**0.8584**	2.7298
	Equiripple	3.62	0.8539	2.7278
San Francisco	Binomial	3.12	0.8170	2.2245
	5/3	**3.11**	**0.9471**	**2.2217**
	9/7	3.12	0.9376	2.2290
	Maxflat	3.12	0.9394	2.2488
	Equiripple	3.12	0.9369	2.2439

The results show that the 5/3 filter is superior to the compared filters in terms of edge sharpness (QE). Another edge quality metric, the NRBM, shows that the edges are less or evenly blurred compared to other filters. It can be also noted that the image *San Francisco* presents better results in all three comparisons, which is due to a high frequency nature of the image. For this type of images, the advantage of 5/3 biorthogonal filter pair can be noticed the best.

Figure 3.14a shows results of 5/3 blending of the *Room* image with two differently exposed halves. The left side of the image is taken from the brighter image, whereas the right side is from the darker. In Fig. 3.14b, 5/3 filter effects are compared to maxflat in the neighborhood of the seam. When the source images have slight brightness difference, 5/3 filter almost completely removes the visible seam (Fig. 3.14b-left), which is not the case with the second-best filter (from Table 3.1 comparison), *maxflat*. The difference is more emphasized in regions where both high and low frequencies are present in the overlapping region. In the cropped part of the *San Francisco* [7] image (Fig. 3.14b-middle), a difference in detail preservation can also be observed in the left, brighter part of the crop. Furthermore, the brightness difference of the background in the left and the right part of the image is slightly reduced.

Figure 3.15 shows luminance values of 50 pixels in one image row around the seam in Fig. 3.14a. Pixel position 0 is the seam. The blue line represents

(a)   (b)

Fig. 3.14 Comparison of different filters used in multi-band image blending. (**a**) Blended image using 5/3 filter. (**b**) *Top row*: blending using 5/3; *Bottom row*: blending using maxflat

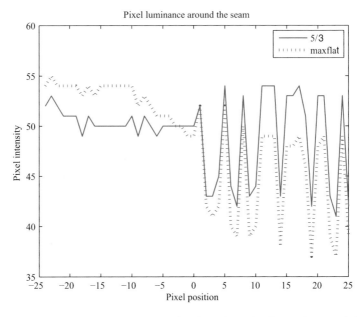

Fig. 3.15 Pixel luminance around the seam using 5/3 and *maxflat* filters. The seam is located at position 0. The 5/3 filters result in smaller intensity difference between the left and the right side of the seam

luminance obtained by 5/3 and the red line represents *maxflat*. Luminance obtained by max is higher left of the seam and lower on the right, compared to 5/3. This is the consequence of brightness levels in initial images. Blending using 5/3 decomposition results in equalized luminance levels around the seam, which are more pleasant for the human visual system.

3.7 Panorama Generation as an Inference Problem

In the previous sections, we explained how to generate a panorama using a known geometry of a calibrated multi-camera systems. The artifacts of the geometric stitching process are reduced using one of the blended methods. However, none of the algorithms take into account the information from the neighboring light rays. Hence, the methodology is still open for development to improve the output image quality.

The panorama generation problem can be defined as a probabilistic inference problem to reduce the seam and parallax errors in the final panoramic image. In [6], the construction of a $360° \times 90°$ FOV panorama from six cameras placed on surfaces of a cubic shape is achieved by mapping the problem on a Markov random field (MRF) and applying belief propagation (BP)-based energy minimization method. Although the system proposed in [6] is composed of a small number of cameras when compared to the one in [2], real-time operation is not reported. Nevertheless, the authors of [6] show that minimizing the energy of the graph provides a final panoramic image that has less ghosting effects, and that the method reduces the errors due to mis-registration of individual image sensors.

Using MRF and Bayesian methods for image processing problems is not new and these methods are well studied in many works [9, 23, 24]. A comparative study is presented in [9] on the utilization of MRFs and applying different energy minimization methods for different vision problems such as object detection, resolution enhancement, and panorama generation. In [24], the belief propagation and graph cut methods are analyzed for energy minimization on MRFs.

In the approach we propose, the real-time panorama generation problem on a multi-camera system is defined as an inference problem, and a methodology is developed to solve it by using MRF for calculating a posterior probability distribution. The method explained in the scope of this book first extracts the priors for generating evidence from the sensor structure and calibration information of the individual cameras, and then generates coefficients from the resulting marginal probabilities. In other words, a posterior probability distribution for each hidden node of MRF is calculated. Then, instead of choosing the best label for the corresponding node, expected value estimation is done for the intensity value of each pixel of the panorama by using the joint probability distributions.

3.7.1 Proposed Approach

The methods for calculation of the intensity values for panorama pixels proposed in [18] and [19] are explained in the previous sections. The three techniques named as the nearest neighbor, alpha blending, and Gaussian weighted blending are given by Eqs. (3.11), (3.12), and (3.13), respectively.

We want to consider the panorama pixel construction problem in the probabilistic domain, unlike the previous methods. The previously proposed blending methods for the spherical multi-camera systems can be classified as follows: The nearest neighbor method given by (3.11) can be seen as choosing the maximum likelihood estimate (MLE) over the probability distribution on the planar ω-plane. The alpha blending expressed by (3.12) can be interpreted as expected value (EV) calculation by using the independent probabilities extracted from the ω-plane projection. Finally, the Gaussian blending given in (3.13) also calculates expected value after redistributing the probabilities by multiplying with Gaussian coefficients. Reconsidering Eq. (3.12), which is a weighted average, the problem of inferring the panorama pixel can be viewed as an expected value calculation:

$$E[X] = \frac{\sum_i p_i x_i}{\sum_i p_i} \tag{3.17}$$

where X is a random variable that can take the value from a set of x_i with a corresponding probability from a set of p_i, and E is the expected value of X. In the same manner, any intensity value in the panorama L can be expressed as an expected value in (3.18) which is equivalent to (3.12).

$$E[L(\mathbf{q}, \boldsymbol{\omega})] = \frac{\sum_i p_i l_i}{\sum_i p_i}, p_i = \frac{1}{|r_i|}, l_i = L(\mathbf{c_i}, \boldsymbol{\omega}) \tag{3.18}$$

The MLE or nearest neighbor chooses one exact pixel captured by one of the cameras. Therefore, at camera transition points on the panorama, visible seams are inevitable and the image is not natural for a human observation as seen in Figs. 3.18a and 3.19a. EV method or linear interpolation makes a combination of the camera pixels by using the distance values on the ω-plane as weights, which is referred as probabilities. It provides smoother transitions but since the probability values are independent of the distribution of the adjacent panorama pixels, it causes ghosting effects at object boundaries.

Finally, EV with Gaussian coefficients (3.13) targets a better distribution for the independent probabilities of each $\boldsymbol{\omega}$ direction regarding to the position of the candidate pixel on the camera image planes. Although the observational choice of Gaussian distribution smoothens the transitions at seams and reduces the intensity differences among the panorama, it does not consider the distribution at neighboring directions. Thus the ghosting effects are not removed completely.

However, by the nature of image reconstruction problem, there is a strong spatial dependency between neighboring pixels. Hence, just applying an independent probability distribution by using projected distances on the planar plane cannot sufficiently deal with the parallax errors and blurring near object boundaries, which is the main driving motivation of this section.

Fig. 3.16 Example graph representation for the panoramic image

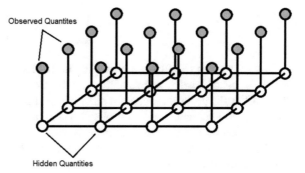

Observed Quantites

Hidden Quantities

3.7.2 Graph Representation

It is proposed to apply a marginalizing process among the probabilities of neighboring ω directions, and to get more accurate probability distribution for each ω direction before calculating the expected intensity value of each panorama pixel. To achieve this, mapping the problem onto an undirected graph where dependency is defined as the connectivity of the nodes is required. Markov random field (MRF) representations are appropriate models for this kind of imaging problems.

The general structure of a MRF is an undirected graph $G = (V,E)$ with observed nodes and hidden nodes as shown in Fig. 3.16. The panoramic image is mapped onto a MRF where the set of nodes V are composed of the hidden nodes X, and the observed random variables Y. The intensity value of a particular direction ω_u is a hidden quantity that will be inferred. The observed values are the intensity values captured by the individual image sensors in the system. Each observed variable has a bias that is calculated as the prior probability value at the ω-plane projection step. The target is to strengthen this bias by considering the neighboring probability distributions such that the resulting panorama pixel intensity will be more natural.

Unlike the many cases where MRFs are used to label each hidden node and choose a best matching random variable, our target is to calculate an expected value for the hidden node by using the probability distribution of the observed variables. Hence, a *maximum a posteriori* (MAP) solution is not taken. Since the final intensity value for each ω_u direction will be calculated by weighted aggregation at the last algorithmic step, we do not take the intensity differences of the neighboring ω_u directions; therefore, a compatibility potential function is not used. Thus, the only potentials used between neighboring hidden nodes X are the ϕ_{iu} as given in (3.19) where u represents a hidden node, i represents the camera indices in the system. So ϕ_{iu} are the probability values of each camera neighboring nodes extracted from the distances on the ω_u-plane. Then the joint probability distribution for each 4-connected neighbor set of a direction ω_u is defined as (3.19). In (3.19), N_u represents the 4-connected neighbor set (i.e., up, down, left, and right neighbors) of direction ω_u where Z is a normalization factor.

$$p_{iu} = \frac{1}{Z}\phi_{iu}\prod_{v\in N_u}\phi_{iv}, \phi_{iu} = \frac{1}{r_{iu}} \tag{3.19}$$

The expression (3.19) is based on the local Markov property that is a variable which is conditionally independent of all other variables given its neighbors. With (3.19), the posterior probability distribution for each node is achieved, which is equivalent to attaining marginal probabilities in the 4-connected neighborhood of a certain direction. Then, these posterior probabilities are replaced as new weights in (3.18) to calculate the expected value as the final intensity for each of the panorama pixel u as in Eq. (3.20)

$$E[L(\mathbf{q}, \boldsymbol{\omega_u})] = \frac{\sum_i p_{iu}l_{iu}}{\sum_i p_{iu}}, l_{iu} = L(\mathbf{c_i}, \boldsymbol{\omega_u}) \tag{3.20}$$

By taking the probability distributions of neighboring nodes into account, stronger evidence is calculated from the initial priors. Hence, the reconstructed panorama is more natural with the reduced defects like ghosting effects and the visible seams. Visual comparisons are provided in experimental results section.

3.7.3 Accurate Prior Estimation from Spherical Model

The planar ω-plane approach as described in Sect. 3.2 and illustrated in Fig. 3.4, which is used for extracting distance values, is based on the constant light flux (CLF) assumption. It is assumed that the light intensity remains constant on the trajectory of any light ray. This assumption is true but not sufficient for the case of the spherical arrangement of the cameras. For a better extraction, the geometry of the camera array should be considered as well. Therefore, the CLF assumption is extended by changing the projection surface to a spherical geometry, which is inspired from the actual physical placements of the individual cameras on the hemispherical frame. This modification is illustrated in Fig. 3.17; the previous projection method is provided in Fig. 3.17a, which can be also extracted from Fig. 3.4. In Fig. 3.17b, where the new proposal is shown, the planar projection surface is replaced by the spherical surface. The light flux is still assumed to be constant but a direction factor is added due to the spherical camera arrangement geometry. According to this new proposal, the light flux is assumed constant on a trajectory of any light ray through the point of observation, which is the center of the sphere in default case.

The virtual spherical surface is constructed by bounding the planar surface with a circle that has a radius of the maximum distance $|r_i|$ for the current ω vector projection. Extending this circle to 3D symmetrically, which results in a hemisphere surface with a radius of r_{imax}. The arc values a_i are defined as the arc distance from the pole of the ω-sphere and the re-projected position of the camera center point

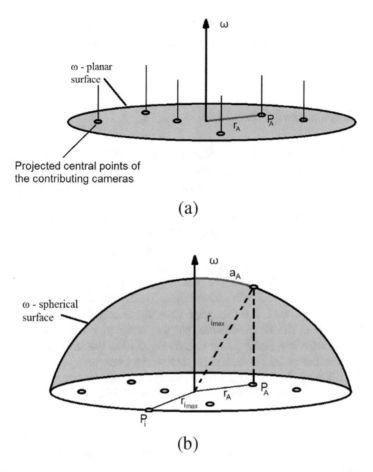

Fig. 3.17 The proposed method for accurate prior estimation from spherical model, (**a**) the previous planar model, (**b**) proposed spherical arrangement for estimating the priors

on the sphere. The relation between the a_i and the corresponding r_i can be solved easily by using simple geometry, which can be expressed in (3.21); and when the distances are normalized to 1, the projection surface becomes a unit sphere. Then, arc distance can be calculated as (3.22).

$$a_i = r_{imax} \left[\frac{\pi}{2} - \arccos \left(\frac{|r_i|}{r_{imax}} \right) \right], \quad r_{imax} = \max \{|r_i|\} \tag{3.21}$$

$$a_{inorm} = \frac{\pi}{2} - \arccos \left(|r_i| \right) \tag{3.22}$$

By applying this transform to the projection surface and the distance values, the absolute values of the relative distances between the projected camera centers on the surface and the vector have become more realistic. That is, further points, which are less likely to be candidate label for the current ω direction, become more distant with respect to the closer points since they move on to the curved surface of the new spherical surface.

3.7.4 Experimental Results

In Fig. 3.18, an example spherical panorama image of a complex scene generated by using the proposed method is provided. The resolution of the final panorama image is 1920×512. The single images are captured by the 24 camera system that is presented in Chap. 5.

The image in Fig. 3.18a is the panorama image of the same scene generated by using the nearest neighbor technique. The comparison of two images in terms of seams can be done on these images. In Fig. 3.18b, the alpha blending result is shown, where the seam is less visible, and the camera boundaries are not as sharp as with the nearest neighbor. However, there are ghosting effects at the camera boundaries especially at object edges. The result of the Gaussian blending is shown in Fig. 3.18c. The object edge ghosting effects and seems are removed when compared to the alpha blending but they are still visible. The result using the proposed inference method is shown in Fig. 3.18d. The improvement is at the camera boundaries and around objects, such as the white polyp in the middle. The zoomed version of the object of interest and the comparisons on the constructed images with the previous methods are shown in Fig. 3.19a, d.

3.8 Inter-Camera Pixel Intensity Differences and Its Applications

There is more than one contributing camera for each virtual ommatidia. In ideal case, if all the cameras are identical with perfect calibration, and if we do not disregard the distances of the objects from the camera, the contributions of each camera for a specific virtual ommatidia direction would be same. The reason is simple; all the cameras should have sampled the same light ray from the surrounding light space for that specific direction.

However, in reality there are nonidealities in various components. As the first nonideality, it is assumed that the objects are sufficiently far from the cameras and the weighted average of the intensity values from different contributing cameras is done. This causes blurry edges and ghosting in the final panoramic image.

Moreover, our calibration method is dependent on the environment, on the distances of the objects at the moment that the calibration takes place, and on the feature points in the scene. The last point is that since the cameras are physically not identical, their color temperature is slightly different from each other. As a result

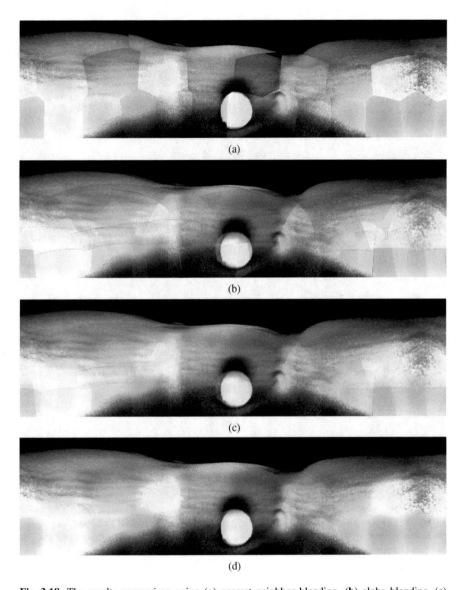

Fig. 3.18 The results comparison using (**a**) nearest neighbor blending, (**b**) alpha blending, (**c**) Gaussian blending, (**d**) the probabilistic inference. The images are 1920×512 pixel resolution

when we add up all these points, there will be intensity differences between the cameras that are supposed to contribute to the same virtual ommatidia direction.

We analyzed this result and utilize these inter-camera differences for different purposes and create a hypothesis for each case.

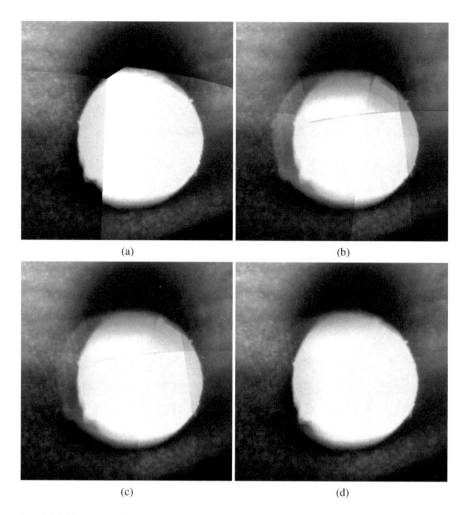

(a)   (b)

(c)   (d)

Fig. 3.19 The comparison of the zoomed polyp region to show the effect on object boundaries: (**a**) nearest neighbor, (**b**) alpha blending, (**c**) Gaussian blending, and (**d**) the proposed probabilistic inference method

3.8.1   Object Boundary Detection

The distance values on the orthogonal plane of virtual ommatidia (VO) directions are used as a measure of how likely a camera can contribute to a virtual ommatidia direction. Smaller distance r_i on the orthogonal plane of ω_j means closer focal vector $\mathbf{t_i}$ of the camera to the vector ω_j. This also means that for the cameras with the larger orthogonal distances, there will be more disparity on the object boundaries, which means higher probability of error. Hence, for a given virtual ommatidia direction ω_j, the pixel intensity difference between the closest and farthest camera results in a quantity that contains information on whether there is a close object in that location. We use the luminance channel of the YCbCr[5] color scheme to compare the pixel intensity difference.

To test this hypothesis, we find the cameras that have the maximum and minimum distances from the observing point center on the orthogonal plane of the VO. The orthogonal plane is illustrated in Fig. 3.17a and the distances are calculated given by the (3.23). The camera indices with minimum and maximum distances are given by (3.24) and (3.25), respectively. The $\mathbf{q}$ is the observation point inside the hemisphere, which is $\{0, 0, 0\}$ by default. Then the object boundary value L_{ob} of each direction is given by (3.26). If we have all the L_{ob} values for all the directions in the compound panoramic image, then we can have the object boundary map of whole field of view.

$$\mathbf{r_i} = (\mathbf{q} - \mathbf{t_i}) - ((\mathbf{q} - \mathbf{t_i}) \cdot \boldsymbol{\omega}) \times \boldsymbol{\omega} \tag{3.23}$$

$$j = \underset{i \in I}{\operatorname{argmin}}(|r_i|) \tag{3.24}$$

$$k = \underset{i \in I}{\operatorname{argmax}}(|r_i|) \tag{3.25}$$

$$L_{ob}(\mathbf{q}, \boldsymbol{\omega}) = \left| L(\mathbf{c_k}, \boldsymbol{\omega}) - L(\mathbf{c_j}, \boldsymbol{\omega}) \right| \tag{3.26}$$

We tested this hypothesis with 15-camera presented in Chap. 4 and with the 24-camera miniaturized system presented in Chap. 5. For example, Fig. 3.20a shows the images that are captured with the system explained in Chap. 4. This system has 15 cameras placed on a hemisphere of 30 mm radius. Each of the cameras is 353×288 pixel resolution (CIF). The panoramic image reconstructed is shown in Fig. 3.20b. The method we proposed [8], which is described in Sect. 3.7.1, is utilized for a panorama generation of 1920×512 pixel resolution. The images are obtained by implementing Eqs. (3.23)–(3.26).

The object boundary map generated by the proposed method is in grayscale as seen in Fig. 3.20a. There are certain regions seen as complete black with $L_{ob} = 0$. These regions are the regions that are seen by only one camera. Hence there is no possible intensity difference value. The regions seen only by one camera are shown in Fig. 3.20c.

In order to have a decision image, a binary image is obtained by applying a threshold value to the object boundary map as given in (3.27). The choice of the ideal threshold value th_{ob} can be made by considering the default intensity value differences, which are not dependent on the scene. As described early in this section, there will be intensity differences due to the calibration imperfections and inter-camera color inconsistency. A resulting image with a threshold value of $th_{ob} = 0.2$ is shown in Fig. 3.20d obtained from the image in Fig. 3.20a where the gray intensity values are in the $[0, 1]$ interval. Here the threshold value is chosen empirically for illustration.

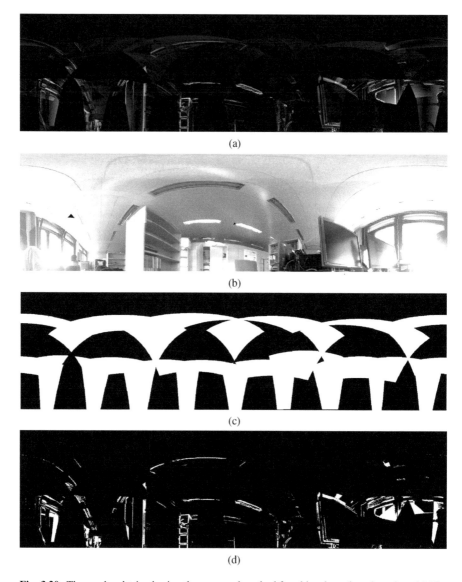

(a)

(b)

(c)

(d)

Fig. 3.20 The results obtained using the proposed method for object boundary detection. (**a**) The object boundary map in grayscale, (**b**) Constructed panoramic image, (**c**) the regions seen by only one camera are shown in *black*, (**d**) the final object boundary map with a threshold $th_{ob} = 0.2$

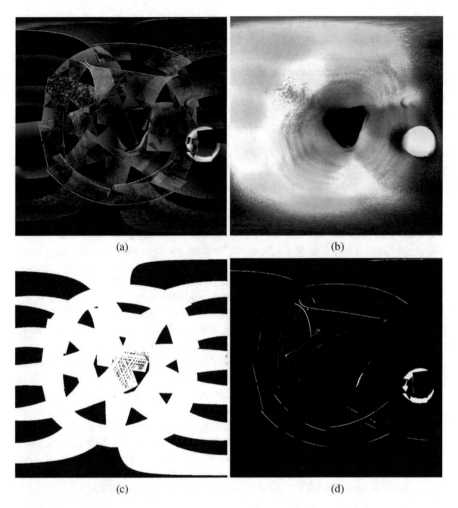

Fig. 3.21 The results obtained from the proposed method for object boundary detection. (**a**) The object boundary map in grayscale, (**b**) constructed panoramic image, (**c**) the regions seen by only one camera are shown in *black*, (**d**) the final object boundary map with a threshold $\text{th}_{\text{ob}} = 0.6$

$$L_{\text{ob}}(\mathbf{q}, \boldsymbol{\omega}) = \begin{cases} 1, & \text{if } L_{\text{ob}} > \text{th}_{\text{ob}}. \\ 0, & \text{otherwise}. \end{cases} \tag{3.27}$$

We also tested the method on the colonoscopy image taken from a human colon model as shown in Fig. 3.21. The images in Fig. 3.21 are $180° \times 180°$ panoramic images with 1080×1080 pixel resolution.

With the visual results, the hypothesis stated in the beginning is tested and validated. That is, for the proposed hemispherical camera system using the pixel mapping method utilized for panorama generation, the inter-camera pixel intensity differences of the most and least confident cameras contain information about the boundaries of the objects. Although this result of the proposed approach is not adequate by itself for an automatic object detection method, it can be extended to machine vision applications such as automatic polyp detection [21].

3.8.2 Inter-Camera Pixel Intensity Differences as Inference Evidences

In the previous panorama construction methods [1, 8, 18, 19], the evidences used as the interpolation weights are purely extracted from the geometrical relations of the cameras under the pinhole camera assumption. The geometrical relations between cameras are fixed and do not change from scene to scene. Therefore, these methods are not taking into consideration the dynamically changing inter-camera pixel intensity differences. We propose to use the camera intensity values as evidences, or weights, for their contribution to the final scene. The ghosting artifacts at the object boundaries can be reduced using this approach.

Before using these differences, a difference metric should be defined. We use the sum of absolute differences (SAD) as the difference measure, because of its simplicity. We again use the luminance channel to calculate the differences.

For each constructed virtual ommatidia direction, the pixel intensity difference of each camera with respect to the other contributing cameras is calculated. Then the total SAD value is calculated for each contributing camera, for each virtual ommatidia direction seen by more than one camera, given by (3.28). We use the reciprocal of the total SAD value as an evidence and combine with the geometrical evidences value by multiplying for each camera in each VO direction as given in (3.29) extending the method described in Sect. 3.7.1. Thus, if the total SAD value is low, it will increase the contribution of the camera to the direction in consideration, and if the SAD value is high, it will decrease its effect.

$$\text{SAD}_i = \sum_{j=1}^{N_{\text{cont}}} \left| I_i - I_j \right| \tag{3.28}$$

$$E\left[L(\mathbf{q}, \boldsymbol{\omega_u})\right] = \frac{\sum_i \frac{1}{\text{SAD}_i} p_{iu} l_{iu}}{\sum_i \frac{1}{\text{SAD}_i} p_{iu}}, \quad l_{iu} = L(\mathbf{c_i}, \boldsymbol{\omega_u}) \tag{3.29}$$

We applied the method to the colonoscopy scene and the resulting image is shown in Fig. 3.22a. The visual result is compared to the resulting image from the method described in Sect. 3.7.1 as shown in Fig. 3.22b. The method performs accurately at the object boundaries for directions seen by more than two cameras since the SAD values are equal and bring equal weights for the directions seen by only two

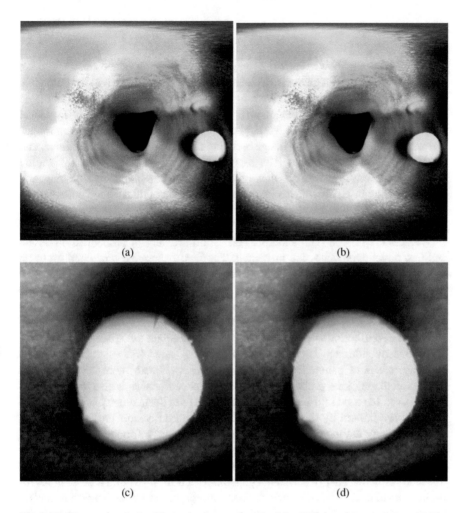

(a)   (b)

(c)   (d)

Fig. 3.22 The results obtained from the proposed method for SAD-based interpolation. (**a**) The resulting image in $180° \times 180°$ angle of view 1080×1080p resolution, (**b**) the same reconstruction using the method from previous section, (**c**) zoomed image on the polyp for SAD-based method, (**d**) zoomed image from the previous method

cameras. A zoomed image comparison for the polyp is given in Fig. 3.22, where a small area on the polyp boundary is seen by three cameras. The resulting image is shown in Fig. 3.22c and the image from the method described in Sect. 3.7.1 in Fig. 3.22d.

3.9 Conclusion

In this chapter, we showed how the image is formed on an image sensor, and we explained how the given geometrical relations can be used to stitch a panorama in a calibrated multi-camera system. We also presented a way to have a programmable FOV, while keeping the same panorama resolution. The differences in image brightness and ghosting are stated as the most noticeable artifacts in panoramas. We presented several real-time image blending methods to overcome this issue, and provided their image examples. The comparison shows that the Gaussian blending results in the improved image quality over the simpler real-time stitching and blending methods. Hence, the hardware implementation of this method will be given in Chap. 4, where the full design of a miniaturized multi-camera system will be presented.

We also considered the panorama generation problem in probabilistic domain and two novel approaches are presented. The final panorama is mapped on to an undirected graph, which is chosen as Markov random field; and the joint probability distributions are calculated on 4-connected neighborhood. Registration information extracted from the camera calibration parameters is used as the likelihood prior and the neighboring pixel probability distributions are used to get more accurate probability distribution in order to get a seamless and natural spherical panorama. The reconstruction step is considered as expected value (EV) estimation for the intensity values of panorama pixels. In order to start with more accurate priors before calculating joint probabilities, a geometrical correction is also proposed.

However, the used registration priors are fixed with the geometry of the structure and they are limited by the mapping method using the vector projection and the pinhole camera assumption. To use the dynamic scene information, the sampled intensity values in the light field can be utilized. The methods described in this chapter for utilizing inter-camera differences are given as example uses. They are intended for better quality panorama generation and for providing a basic input for further machine vision tasks such as automatic polyp detection. To accommodate the methods in full performance, systems that have more FOV overlap are desired.

References

1. Afshari H, Akin A, Popovic V, Schmid A, Leblebici Y (2012) Real-time FPGA implementation of linear blending vision reconstruction algorithm using a spherical light field camera. In: IEEE workshop on signal processing systems, pp 49–54. doi:10.1109/SiPS.2012.49
2. Afshari H, Popovic V, Tasci T, Schmid A, Leblebici Y (2012) A spherical multi-camera system with real-time omnidirectional video acquisition capability. IEEE Trans Consum Electron 58(4):1110–1118. doi:10.1109/TCE.2012.6414975
3. Bouget J (2015) Camera calibration toolbox for MATLAB. http://www.vision.caltech.edu/bouguetj/calib_doc. Accessed 20 Oct 2015
4. Brown M, Lowe D (2007) Automatic panoramic image stitching using invariant features. Int J Comput Vis 74(1):59–73

5. Brown W, Shepherd BJ (1994) Graphics file formats; reference and guide. Manning Publications Co., Greenwich

6. Brunton A, Shu C (2006) Belief propagation for panorama generation. In: Third international symposium on 3D data processing, visualization, and transmission

7. Burt P, Adelson E (1983) A multiresolution spline with application to image Mosaics. ACM Trans Graph 2(4):217–236. doi:10.1145/245.247

8. Cogal O, Popovic V, Leblebici Y (2014) Spherical panorama construction using multi sensor registration priors and its real-time hardware. In: IEEE international symposium on multimedia (ISM). IEEE, Washington

9. Felzenszwalb PF, Huttenlocher DP (2006) Efficient belief propagation for early vision. Int J Comput Vis 70(1):41–54

10. Goshtasby A (2014) Multi-exposure image dataset. http://www.imgfsr.com/. Accessed 12 Feb 2014

11. Hartley RI, Zisserman A (2004) Multiple view geometry in computer vision, 2nd edn. Cambridge University Press, Cambridge

12. Kang SB, Weiss RS (2000) Can we calibrate a camera using an image of a flat, textureless lambertian surface? In: Proceedings of the 6th European conference on computer vision - part II, pp 640–653

13. Levoy M, Hanrahan P (1996) Light field rendering. In: Proceedings of the 23rd annual conference on computer graphics and interactive techniques, New York (SIGGRAPH '96), pp 31–42. http://doi.acm.org/10.1145/237170.237199

14. Marziliano P, Dufaux F, Winkler S, Ebrahimi T (2002) A no-reference perceptual blur metric. In: Proceedings of international conference on image processing, vol 3, pp 57–60. doi:10.1109/ICIP.2002.1038902

15. Mittal A, Soundararajan R, Bovik AC (2013) Making a completely blind image quality analyzer. IEEE Signal Process Lett 22(3):209–212

16. Peleg S, Herman J (1997) Panoramic mosaics by manifold projection. In: IEEE conference on computer vision and pattern recognition, San Juan, Puerto Rico, pp 338–343. doi:10.1109/CVPR.1997.609346

17. Popovic V, Leblebici Y (2015) FIR filters for hardware-based real-time multi-band image blending. In: Proc. SPIE 9400, real-time image and video processing, 94000D, San Francisco. doi:10.1117/12.2078889

18. Popovic V, Afshari H, Schmid A, Leblebici Y (2013) Real-time implementation of Gaussian image blending in a spherical light field camera. In: Proceedings of IEEE international conference on industrial technology, pp 1173–1178. doi:10.1109/ICIT.2013.6505839

19. Popovic V, Seyid K, Akin A, Cogal O, Afshari H, Schmid A, Leblebici Y (2014) Image blending in a high frame rate FPGA-based multi-camera system. J Signal Process Syst 76:169–184. doi:10.1007/s11265-013-0858-8

20. Sonka M, Hlavac V, Boyle R (2008) Image processing, analysis, and machine vision, 3rd edn. Cengage Learning, Boston

21. Stanek SR, Tavanapong W, Wong J, Oh JH, De Groen PC (2012) Automatic real-time detection of endoscopic procedures using temporal features. Comput Methods Prog Biomed 108(2):524–535

22. Szeliski R (2011) Computer vision: algorithms and applications. Springer, New York. doi:10.1007/978-1-84882-935-0

23. Szeliski R, Zabih R, Scharstein D, Veksler O, Kolmogorov V, Agarwala A, Tappen M, Rother C (2008) A comparative study of energy minimization methods for Markov random fields with smoothness-based priors. IEEE Trans Pattern Anal Mach Intell 30(6):1068–1080

24. Tappen MF, Freeman WT (2003) Comparison of graph cuts with belief propagation for stereo, using identical MRF parameters. In: Proceedings of the Ninth IEEE international conference on computer vision, 2003. IEEE, Washington, pp 900–906

25. Xydeas C, Petrović V (2000) Objective image fusion performance measure. Electron Lett 36(4):308–309. doi:10.1049/el:20000267

Chapter 4
Omnidirectional Multi-Camera Systems Design

This chapter explains the Panoptic camera, a real-time omnidirectional multi-camera system. The system is composed of a custom printed circuit board (PCB), with the full FPGA-based processing system on it. We explain the full processing pipeline, starting with the motivation and the system constraints for building such a system. It is followed by the system architecture and block-by-block implementation details. We finish with discussion of the experimental results.

4.1 Introduction

In Chap. 2 we gave an overview of the currently available omnidirectional imaging systems. Majority of these systems provide very good image quality at the expense of speed. Whereas this trade-off is acceptable for photographers, applications such as autonomous navigation, telepresence, remote monitoring, or object tracking are strongly dependant on real-time processing. State-of-the-art systems for these applications are currently relying on a single camera solutions, due to lack of fast processing hardware on the market. The multi-camera systems provide omnidirectional capability, and they would certainly improve performance of the navigation or monitoring systems.

Multiple image stitching and blending is the most process intensive part of generating the image mosaic. They are usually implemented as a post-processing operation on a PC. Real-time operation is a very challenging problem. Hence, a GPU implementation or a dedicated hardware solution is often considered. Various existing GPU implementations of image mosaicing algorithms [5, 8] exist, but the problem with GPU implementations is in the scalability and the limited bandwidth for data transfer from the camera system to the PC. On the other hand, FPGAs are widespread used platforms, that enable fast development and prototyping. The important advantages of FPGAs over GPUs are easier performance optimization

© Springer International Publishing AG 2017
V. Popovic et al., *Design and Implementation of Real-Time Multi-Sensor Vision Systems*, DOI 10.1007/978-3-319-59057-8_4

and scalable designs. Processing speed of an FPGA system linearly drops with the increase of the image resolution, making the system scaling predictable. Considering these results, an FPGA-based system is chosen as the processing platform for the Panoptic camera.

Three main parts can be identified in Panoptic camera:

1. Image acquisition—Consisting of the image sensor and interfacing hardware
2. Image processing unit—An FPGA-based system implementing the omnidirectional vision reconstruction
3. User interface—Camera control and panoramic image display.

4.2   Image Acquisition Module

The main component of the image acquisition module is the commercial-off-the-shelf (COTS) image sensor. Since the portability of the system is desired, one criterion for the imager selection is its size. The second criterion is the camera's embedded image processing system-on-chip (SoC), which implements the fundamental raw image processing pipeline. Most of the COTS modules designed for mobile phone integration meet these criteria, and we chose a low-cost PIXELPLUS PO4010N image sensor. The summary of the electrical characteristics of this sensor is given in Table 4.1. Figure 4.1a, b shows the image sensor with the connector, and the hemispherical dome where the sensors should be mounted.

Figure 4.1c illustrates the internal block diagram of PO4010N sensor. The Image Signal Processing block is dedicated to raw image processing, e.g., Bayer-to-RGB conversion, white balancing, auto exposure, etc. It is controlled via Inter-Integrated Circuit (I2C) bus. The sensor outputs 8-bit parallel data, horizontal and vertical synchronization control signals, and the pixel clock used for synchronous transfer.

Table 4.1 Main PO4010N characteristics

Parameter	Value
Total pixel array	386×320
Effective pixel array	368×304
Pixel size	$3.6\,\mu m \times 3.6\,\mu m$
Filter	RGB Bayer color filter
Data interface	8-Bit parallel + synchronization + clock
Frame rate	25 fps
Maximum output resolution	352×288 (CIF)

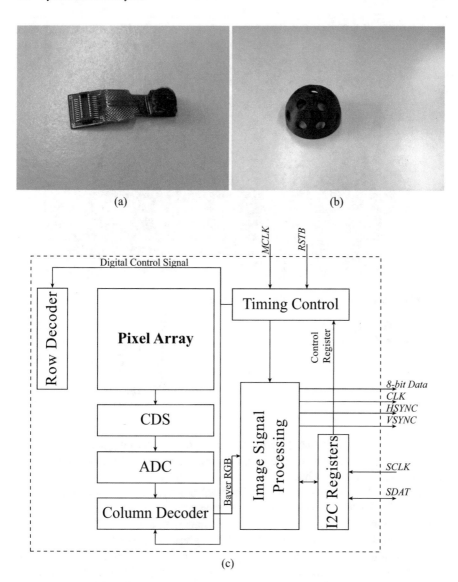

Fig. 4.1 (**a**) PIXELPLUS PO4010N image sensor module, (**b**) the hemispherical dome for sensor placement, and (**c**) block diagram of PO4010N internal architecture

4.3   System-Level Analysis

The state-of-the-art multi-camera systems are built using COTS modules placed in an array, grid, or circular formation. The direct implementation of a multi-camera system consists of connecting all the imaging devices to a central processing unit in a star topology formation. The responsibilities of the central unit are controlling

Fig. 4.2 Virtex-5 processing board for Panoptic camera

Table 4.2 Summary of the available resource on Virtex-5 XC5VLX50

Resources	Number
Logic slices	7200
Block RAM capacity	1728 Kb
DSP blocks	48
User I/Os	560

the cameras, receiving the video streams, storing them into the memory, and implementing the processing algorithm for real-time panorama construction. This is called a *centralized* architecture.

A custom FPGA board (Fig. 4.2) has been designed using a XILINX Virtex-5 XC5VLX50-1FF1153C FPGA as a core processing unit in order to capture and process the video streams produced by the cameras in real-time. The summary of the available resources on this is FPGA chip is given in Table 4.2.

This board directly interfaces with at most twenty PIXELPLUS PO4010N cameras, forming the aforementioned centralized architecture. The number of cameras connected to a single board is limited by the user I/O pin availability of the chosen FPGA chip. To support higher number of camera interfaces, multiple identical boards of the same kind can be stacked. The camera modules provide output data in 16-bit RGB format with selectable frame rate. The 16-bit pixels are in RGB565 format, i.e., red channel is coded with five bits, green with six, and blue with five. The cameras of the Panoptic System are calibrated for their true geometrical position in the world space, their intrinsic parameters, and the lens distortion coefficients [2]. Even though the camera calibration is precise within certain error bounds, the spherical arrangement of the cameras, i.e., diverging camera directions, emphasize parallax problems. Hence, appropriate blending algorithms are still needed for a seamless and ghost-free panorama.

4.3.1 System Memory and Bandwidth Constraints

The incoming video streams from the cameras have to be stored in the memory, and then later processed. The memory is segmented into $N_{cam} = 20$ segments of $C_w \times C_h \times$ BPP bits, where C_w and C_h are frame width and height, and BPP is the number of bits used for representation of one pixel. The needed memory capacity is:

$$M_{cap} \geq N_{cam} \times C_w \times C_h \times BPP \qquad (4.1)$$

Another constraint related to the memory is its bandwidth. The bandwidth of the memory should sustain the aggregate number of access times for writing all the image sensor video streams and that of the application process within the defined fps timing limit. The required memory bandwidth to support real-time storage of frames for N_{cam} cameras is:

$$M_{BW} \geq FR \times N_{cam} \times C_w \times C_h \times BPP \qquad (4.2)$$

where FR is the video frame rate in fps.

The required bandwidth for two different real-time frame rates and six different standard camera resolutions is shown in Table 4.3. The bandwidth is calculated per camera and for BPP = 16 bits. As observed in Table 4.3, an increase in the camera resolution results in large bandwidth requirement. For example, a video stream from the state-of-the-art 20 Mpixels sensor occupies almost full bandwidth of the latest DDR3 memories.

Since twenty PO4010N cameras are envisioned for this board, the system must be able to store their video streams. Multiplying values from Tables 4.3 and 4.4 by 20, the required memory bandwidth and capacity are:

$$M_{BW} \geq 810\,\text{Mb/s}$$
$$M_{cap} \geq 32.4\,\text{Mb} \qquad (4.3)$$

for the frame rate of 25 fps. Hence, two Zero Bus Turnaround (ZBT) Static Random Access Memories (SRAM) with 36 Mb capacity and an operating bandwidth of

Table 4.3 Required memory bandwidth per camera in Mb/s

Camera resolution		FR (fps)	
C_w	C_h	25	30
320	240	30.5	36.6
352	288	40.5	48.6
640	480	122.5	147
1024	768	314.5	377.4
1920	1080	829.4	995.3
5120	3840	7864.3	9437.2

Table 4.4 Required memory space per camera in Mb

Camera resolution		Memory space (Mb)
C_w	C_h	
320	240	1.23
352	288	1.62
640	480	4.92
1024	768	12.59
1920	1080	33.17
5120	3840	314.57

2.67 Gb/s are used on this FPGA board. One memory chip is used for storing the incoming image frames from twenty cameras, while the previous frame is fetched by the image processing core from the other memory chip. The two chips swap their role with the arrival of each new frame.

4.4 Top-Level Architecture

The architecture of the FPGA is depicted in Fig. 4.3. The FPGA design consists of five major blocks. The arrow lines depicted in Fig. 4.3 show the flow of image data inside the FPGA. Image data streaming from the cameras enters the FPGA via the camera interface block. A time-multiplexing mechanism is implemented to store the incoming frame data from all the camera modules into one of the single-port SRAMs. Hence, the camera multiplexer block time-multiplexes the data received by the camera interface block and transfers it to the memory controller for storage in one of the ZBT SRAMs. The SRAMs are partitioned into twenty equal segments, one for each camera. The memory controller interfaces with two external SRAMs available on the board at the same time. The memory controller block provides access for storing/retrieving the incoming/previous frame in/from the SRAMs. As mentioned in the previous section, the SRAMs swap their roles (i.e., one is used for writing and one for reading) with the arrival of each new image frame from the cameras.

The image processing and application unit is in charge of the image processing and basic camera functionalities, such as single video channel streaming, or all-channel image capture. This block accesses the SRAMs via the memory controller, processes the fetched pixels according to the chose application mode, and transfers the processed image to the data link controller. The data link controller provides transmission capability over the external interfaces available on the board, such as the USB 2.0 link. Finally, the camera controller block is in charge of programming and synchronizing the cameras connected to the FPGA board via I^2C bus.

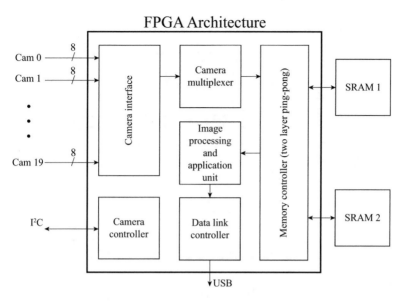

Fig. 4.3 Top-level architecture of the Panoptic camera FPGA design. Arrows denote the data flow

4.5 Implementation of the Image Processing Unit

The panorama construction algorithm is implemented inside the image processing and application unit. The block diagram is shown in Fig. 4.4. This image processing entity comprises five modules, which are thoroughly discussed in the following sections.

4.5.1 Angle and Omega Vector Generation

The angle generation module generates the spherical coordinates $(\theta_\omega, \phi_\omega)$ of the ω directions which are of interest for the reconstruction. It has the ability of generating angles for both equiangular and constant pixel density pixelization schemes from (3.6) and (3.8). The span and resolution of the output view is selectable within this module. It is possible to reconstruct a smaller portion of the light field with an increased resolution, due to the initial oversampling of the light field, i.e., the cameras record more pixels than displayed in the reconstructed image, as explained in Sect. 3.2.2. Hence, a more detailed image with a limited FOV can be reconstructed while keeping the same frame rate. Furthermore, higher resolutions can be achieved by trading-off the frame rate. The 13-bit representation of the angles leaves enough margin for a truthful reconstruction, considering the used CIF imagers and the total amount of the acquired pixels. The generic N-bit representation is as follows:

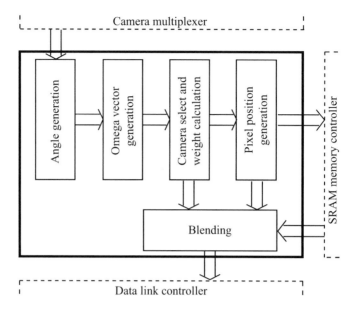

Fig. 4.4 Block diagram of the image processing and application block dedicated to the panorama construction. Each of the blocks within the *bold rectangle* is explained in detail in this chapter

$$b_0 b_1 \cdots b_{N-1} = 2\pi \sum_{i=0}^{N-1} b_i \cdot 2^{-(i+1)} \tag{4.4}$$

Since the coordinate angles are represented by 13 bits, the maximum reconstruction resolution for a hemisphere is 16 Mpixels.

The angles θ_ω and ϕ_ω are implementable using an accumulator for each angle. To generate all possible combinations of θ_ω and ϕ_ω, the θ_ω accumulator increments when the ϕ_ω accumulator completes its full range cycle. This concept is shown in Fig. 4.5a. The two accumulators can have different incrementing steps K_ϕ and K_θ. These incrementing steps define the resolution of the constructed omnidirectional vision. In addition, the limits $\phi_{\min}$, $\phi_{\max}$, $\theta_{\min}$, and $\theta_{\max}$ are set for both angles. The FOV of the constructed panorama is determined through the configuration of these four parameters.

The Omega vector generation module calculates the radial unit vector pertaining to the spherical position $(\theta_\omega, \phi_\omega)$ received from the angle generation module. The vectors are generated according to the following equation:

$$\omega = \sin \theta_\omega \cos \phi_\omega \, x + \sin \theta_\omega \sin \phi_\omega \, y + \cos \theta_\omega \, z \tag{4.5}$$

Trigonometric functions $\sin(2\pi x)$ and $\cos(2\pi x)$ are used for the calculation of the ω vector from the ϕ_ω and θ_ω angles. The implementation of the trigonometric functions sine and cosine has been the focus of direct digital frequency synthesizers

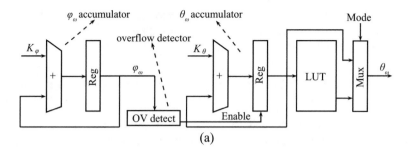

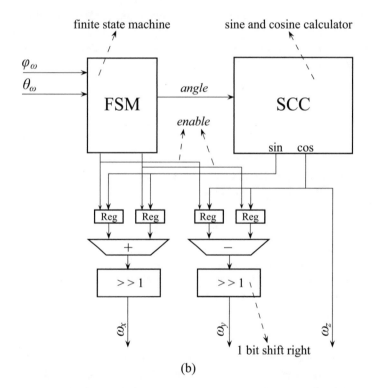

Fig. 4.5 (**a**) ϕ_ω and θ_ω angle generation hardware. The LUT at the output stores the θ_ω values for the equal density discretization; (**b**) hardware implementation of the ω generation block

(DDFS) for the past decades. Hence, many algorithms have been developed for the purpose of the implementation of basic trigonometric functions. Look-up table (LUT)-based algorithms [6], the CORDIC algorithm [7], and polynomial approximation-based algorithms [4] are three widely accepted implementations. The LUT methods are the fastest and numerically the most precise, but they need memory storage. Since there is enough available BlockRAMs in the FPGA (Table 4.2), the LUT method is implemented in Panoptic camera.

The multiplication operations in (4.5) are replaced with their addition-based identities to reduce the number of needed multipliers, since their availability is limited in this FPGA:

$$\boldsymbol{\omega} = \frac{1}{2}(\sin(\theta_\omega + \phi_\omega) + \sin(\theta_\omega - \phi_\omega))\mathbf{x} + \frac{1}{2}(\cos(\theta_\omega - \phi_\omega) - \cos(\theta_\omega + \phi_\omega))\mathbf{y} + \cos(\theta_\omega)\mathbf{z}$$

$$(4.6)$$

The $\mathbf{x}$, $\mathbf{y}$, and $\mathbf{z}$ components of the $\boldsymbol{\omega}$ are calculated utilizing a finite state machine (FSM), which provides three angles from (4.6) consecutively, and a single SCC module. The SCC module calculates and outputs the sine and cosine values simultaneously. Hence, the $\boldsymbol{\omega}$ is obtained by presenting the following angle combinations to the SCC module: $(\theta_\omega + \phi_\omega)$, $(\theta_\omega - \phi_\omega)$, and θ_ω, and combining their respective sine and cosine outputs in the correct order. An architectural view of the $\boldsymbol{\omega}$ vector generation module is shown in Fig. 4.5b.

4.5.2 Camera Selection and Weight Calculation

The camera select and weight calculation module determines which cameras contribute to the construction of the pixel in $\boldsymbol{\omega}$ direction. The camera c_i is considered to be contributing if the $\boldsymbol{\omega}_j$ is within it FOV α_i, i.e.,

$$\omega_t = \boldsymbol{\omega}_j \cdot \mathbf{t_i} > \cos\left(\frac{\alpha_i}{2}\right)$$

$$(4.7)$$

Furthermore, this module computes the distance between the focal point projection and the virtual observer projection on the ω-plane, for each contributing camera c_i in direction $\boldsymbol{\omega}_j$, as expressed in (4.8):

$$r_{i,j} = |(\mathbf{q} - \mathbf{t_i}) - ((\mathbf{q} - \mathbf{t_i}) \cdot \boldsymbol{\omega}_j)\,\boldsymbol{\omega}_j|$$

$$(4.8)$$

The pseudo-code describing the operation of this module is given in Algorithm 1, and its hardware implementation in Fig. 4.6. For each pixel direction $\boldsymbol{\omega}$ that is received from the Omega vector generation module, this module sequentially provides the stored $\mathbf{t}$ values of all interfaced cameras, and calculates the ω_t dot product. The condition from (4.7) is checked in the Minimum Search sub-block. This architecture ensures consecutive checks of all Panoptic interfaced cameras, and their contribution to the selected $\boldsymbol{\omega}$ direction.

Concurrently, the distance vectors of the projected focal points of the cameras with the projected virtual observer point are calculated (4.8). The magnitude of the $\mathbf{r}$ is passed as the distance value to the Minimum Search sub-block. If the NN blending is chosen, Minimum Search sub-block finds the minimum distance r, and selects the corresponding camera as the only one at the output. If alpha or

---

Algorithm 1 Camera select and weight calculation

1: $r_{min} \leftarrow 1$
2: **for all** cameras **do**
3:      $\omega_t \leftarrow \boldsymbol{\omega} \cdot \mathbf{t}$
4:      $\mathbf{r} \leftarrow (\mathbf{q} - \mathbf{t}) - ((\mathbf{q} - \mathbf{t}) \cdot \boldsymbol{\omega}) \, \boldsymbol{\omega}$
5:      **if** $(\omega_t > \cos(\frac{\alpha}{2}))$ **then**
6:          **if** blending $==$ nearest_neighbor **then**
7:              **if** $(|\mathbf{r}| < r_{min})$ **then**
8:                  $r_{min} \leftarrow |\mathbf{r}|$
9:                  STORE camera_index
10:             **end if**
11:         **else**
12:             $r \leftarrow |\mathbf{r}|$
13:             index $\leftarrow$ camera_index
14:         **end if**
15:     **end if**
16: **end for**
17: **if** blending $==$ nearest_neighbour **then**
18:     $r \leftarrow r_{min}$
19:     index $\leftarrow$ camera_index
20: **end if**

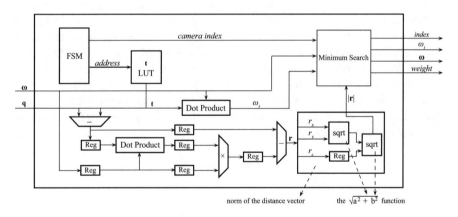

Fig. 4.6 Block diagram of the camera select and weight calculation module

Gaussian blending is selected, the search is not necessary, i.e., indices and weights of all contributing cameras are provided at the output, sequentially. The weights are calculated using (3.12) and (3.13).

4.5.2.1   Dot Product and Square Root Sub-blocks

The dot product calculator of this module is implemented using a pipelined architecture consisting of three multipliers and two adders. Three multipliers are followed by a register stage. Two products are added together, whereas the third is again registered. Finally, the dot product is obtained after the final addition that is again registered. The pipelined architecture is used to boost the system performance and shorten the critical path that would be very long if both a multiplier and an adder were on it.

A norm of a 3D vector is also calculated using the pipelined architecture, by cascading two 2D vector norm calculators, i.e., $\sqrt{a^2 + b^2}$. The 2D norm calculator uses CORDIC algorithm, and it is implemented using only addition and subtraction operators [7].

4.5.3 Pixel Position Generation

The pixel position generation module calculates the true pixel position in the image frame of the cameras selected in the camera selection block. This goal is achieved using the pinhole camera model [3] to obtain the two-dimensional position (d_x, d_y) in the camera image plane, which is identified by the vectors **u** and **v**. The basic pinhole camera model equation is expressed in (3.5).

In reality, the mapping of a 3D scene onto an observed 2D plane of a camera image frame is a complex problem, which is only coarsely represented by (3.5). The intrinsic parameters of the camera are categorized in two classes, and characterize the mapping between a 3D scene and the observed 2D plane. The first class is the linear homography, defined by a 3×4 camera matrix, mapping of 3D points coordinates into 2D pixel coordinates [3]. The second class models the non-linear effects such as lens distortion. These parameters are estimated through the calibration process [2], stored for each camera in a LUT, and applied to results of the basic pinhole camera model as expressed in (4.11):

$$\omega_u = \boldsymbol{\omega} \cdot \mathbf{u}$$
$$\omega_v = \boldsymbol{\omega} \cdot \mathbf{v} \tag{4.9}$$

$$d_x = f \cdot \frac{\omega_v}{\omega_t}$$
$$d_y = f \cdot \frac{\omega_u}{\omega_t} \tag{4.10}$$

$$R^2 = d_x^2 + d_y^2$$
$$\text{poly} = k_5 R^6 + k_2 R^4 + k_1 R^2 + 1$$
$$d_x' = \text{poly} \cdot d_x + 2k_3 d_x d_y + k_4 (R^2 + 2d_x^2)$$
$$d_y' = \text{poly} \cdot d_y + 2k_4 d_x d_y + k_3 (R^2 + 2d_y^2) \tag{4.11}$$

The obtained (d_x', d_y') values do not necessarily match with exact pixel positions in the camera frame. The nearest pixel position is chosen by simply rounding the

(d'_x, d'_y) values, and getting the true (x, y) coordinates. Hence, the intensity value of the closest pixel (x, y) is chosen as the intensity value of (d'_x, d'_y). The same process is repeated for each camera frame, i.e., for all $\mathbf{u}$ and $\mathbf{v}$ vectors.

The pixel position generation module interfaces with the SRAM memory controller to retrieve the pixel value of the contributing cameras upon calculation of the true pixel position. The camera index originating from the camera select module is used to access the correct memory segment of the SRAM, i.e., the segment where the image frame of the selected camera is stored. The true (x, y) coordinates are mapped to their corresponding addresses in the memory segment used for storing the camera frame.

A detailed block diagram of this module is shown in Fig. 4.7. The proposed architecture is pipelined, which allows streaming access to the true pixel positions. The pipeline registers are omitted for clarity purpose.

The pixel position module also calculates the distance of the selected pixel in the image frame from the image center. This distance is represented as R' in Fig. 4.7 and it is further used for Gaussian blending and the vignetting correction.

4.5.3.1 Sub-blocks of Pixel Position Generator

Hardware dedicated to the lens distortion compensation is marked with a box in Fig. 4.7. Distortion compensation of the cameras is conducted on-the-fly in the proposed architecture rather than through the large transformation LUT for each camera, which demands additional memory space. A resource-efficient architecture has been achieved through factorization of similar terms, as shown in (4.10).

The arithmetic division operators are implemented using a four-stage iterative convergence method described in [1]. The polynomial function $f(R^2)$ that calculates the poly term in (4.11) is implemented using the Horner scheme [4] implementation of the polynomial functions. This method achieves higher precision in less number of iterations compared to the other alternative methods [4].

4.5.4 Image Blending

The blending module conducts the final step of panorama construction. Three blending methods are currently available on Panoptic camera: nearest neighbor, alpha blending, and Gaussian blending.

The module receives the pixel values from all contributing cameras along with the blending weight (except when NN blending is selected), and the distance R', as shown in Sects. 4.5.2 and 4.5.3. The weights are used to calculate the final pixel value using one of Eqs. (3.11), (3.12), or (3.13). As the final result, the block estimates a single light intensity value for each ω direction.

A straightforward implementation of the blending algorithms is very resource-demanding. Multipliers and dividers would be required for each color channel

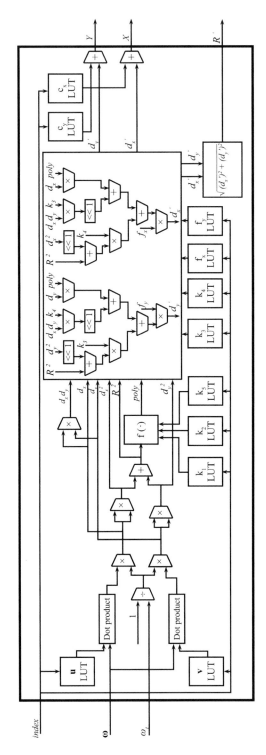

Fig. 4.7 Block diagram of the pixel position generation module. The pipeline registers are omitted for clarity

separately. Thus, it is beneficial to share hardware resources which are common for all channels, e.g., implementation of the denominator in (3.13). For the implementation purpose, (3.13) is expressed differently:

$$\mathscr{L}(\mathbf{q}, \boldsymbol{\omega}) = \sum_{i \in I} a_i \cdot \mathscr{L}(c_i, \boldsymbol{\omega})$$

$$a_i = \frac{w_i \cdot G(R_i')}{\sum_{k \in I} w_k} \tag{4.12}$$

where a_i is the normalized weight of the ith contributing camera, and $G(R_i')$ is the vignetting correction factor. The vignetting correction factors are pre-calculated and stored in a LUT. As explained in Sect. 3.3, this factor is addressed in LUT using only the pixel distance from the camera's optical center. The pseudo-code of the blending block and its hardware implementation are given in Algorithm 2 and Fig. 4.8. I_{RGB} represents color intensities of the contributing pixels in the algorithm notation.

Algorithm 2 Blending

1: **if** blending == nearest_neighbor **then**
2:    $\mathscr{L}_{RGB} \leftarrow I_{RGB}$
3: **else**
4:    $w_{acc} \leftarrow \sum_{k \in I} w_k$
5:    **for all** $i \in I$ **do**
6:       $a_i \leftarrow w_i \cdot \frac{1}{w_{acc}}$
7:       $a_i \leftarrow a_i \cdot G(R_i')$
8:    **end for**
9:    **for all** color_channels **do**
10:       $\mathscr{L}_{RGB} \leftarrow \sum_{i \in I} I_{RGB} \cdot a_i$
11:    **end for**
12: **end if**

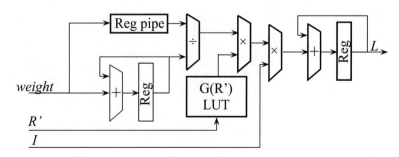

Fig. 4.8 Block diagram of the image blending module

4.6   Experimental Results of the Panoptic System

A Panoptic hemisphere of diameter $2r = 3$ cm is built using a 3D printer. It can accommodate fifteen PO4010N cameras, arranged on three floors. The hemisphere populated with cameras is positioned on top of a plexiglas structure that is attached to the designed PCB, as shown in Fig. 4.9. The camera modules are connected to the digital interfaces on the PCB using flexible PCBs, and they are operated at 25 fps.

The architecture presented in this chapter was developed in VHDL for the target FPGA. The developed firmware conducts all mathematical processing using 16-bit fixed-point precision. The sufficient bit precision of the mathematical operations was verified through the MATLAB model. The system control unit shown in Fig. 4.3 is implemented using a Xilinx Picoblaze soft-core 8-bit microcontroller.

The firmware was targeted and successfully tested for operation at 133 MHz f_{clk} frequency. The total latency of the system is 132 clock cycles, which is less than 1 μs, using 133 MHz frequency. The power consumption of the FPGA board, when its firmware is in full operation, was measured at 5.05 W. The aspect ratio of the reconstructed panorama is kept at 1:4 to be support $90° \times 360°$ FOV without any additional distortions. Hence, the firmware is able of constructing an omnidirectional view with the resolution of 0.25 MPix (256×1024) at the rate of:

$$\text{FR} = \frac{f_{clk}}{N_{cam} \cdot C_w \cdot C_h} = 25.3 \text{ fps} \qquad (4.13)$$

which is the frame rate of the cameras providing the input video streams.

The blending methods presented Sects. 3.4 and 3.5 were separately implemented on the FPGA in order to compare their resource utilization. The summary is presented in Table 4.5. Gaussian and Adaptive Gaussian blending infer additional

Fig. 4.9 The fully assembled Panoptic camera system

Table 4.5 Panoptic Virtex-5 FPGA resource utilization comparison

Blending	Nearest neighbor	Linear	Gaussian	Adaptive Gaussian	
Resource	Used				Available
Slices	4070	4653	4607	4816	7200
Slice registers	9351	10,069	10,127	10,196	28,800
BlockRAMs	17	17	21	22	48
DSPs	37	47	48	48	48

LUTs and multipliers compared to NN and alpha blending. This is observed through the increase of the used BlockRAMs and logic slices. However, the increase of resource usage compared to the alpha blending and NN is very small and is not an influential factor in the overall utilization.

Four image captures from real-time panoramic reconstruction are shown in Fig. 4.10. The Gaussian blending with $\sigma_d = 100$ was used.

4.7 User Interface and Display

The external host for the developed prototype is a PC that interfaces with the developed FPGA platform. The role of the PC is restricted to video display and parameter selection in the FPGA firmware. No additional image processing is conducted on the PC side. A Graphical User Interface (GUI) is developed in the PC for controlling the Panoptic camera. The GUI also offers video display, image capture, and recording capability.

The Panoptic camera can be used as a perfect example of a telepresence system. Unlike the virtual reality systems, where users are transported to a virtual scene, telepresence allows users to be in another location in the real world, e.g., videoconferencing. Among the benefits of videoconferencing, we can say it lowers the travel requirements, improves dialog efficiency, and allows the mobility-impaired people to visit distant places. Instead of using the narrow FOV cameras, we can achieve a better telepresence experience.

The setup used to test and demonstrate the system consists of the Panoptic camera, a PC, and the virtual reality headset Oculus Rift HMD. The camera generates omnidirectional images at a resolution of 1024×256 pixels and transmits them via USB 2.0 link to the PC.

The developed GUI supports Oculus Rift, and creates a virtual environment in order to display the hemispherical image. This virtual environment is created using the OpenGL API and consists of a user controlled camera and a large overhead hemisphere, onto which the image is projected. The omnidirectional image is used as a texture for the virtual hemisphere. The camera rotates according to the sensor data received from the head-mounted display.

(a)

(b)

(c)

(d)

Fig. 4.10 Image captures from the Panoptic video stream. The Gaussian blending with $\sigma_d = 100$ is used in all images. The scenes are from different locations around EPFL campus

Figure 4.11a shows the textured virtual hemisphere from the side. When using the application with the head-mounted display, the user viewpoint is in the middle of the sphere. Figure 4.11b shows the application in normal use with the HMD.

In order to ensure a high frame rate at all times, the application receives new omnidirectional images in a secondary thread. Thanks to the multi-threaded implementation, the rendering frame rate is independent from the USB speed, as well as the camera frame rate. This is important for the streaming functionality, in which the frame rate can vary.

(a) (b)

Fig. 4.11 (**a**) The textured OpenGL hemisphere showing a captured image, viewed from the side. (**b**) The client application generating the left- and right-eye view for the head-mounted display

4.8 Conclusion

The current trend in building multi-camera systems is to use multiple commercially available camera modules. Panoptic, a real-time multi-camera system is presented in this chapter. Panoptic is implemented using the centralized processing approach of the multi-camera system. The full design was detailed, including the camera choice and specifications, constraint analysis, a real-time hardware implementation of the omnidirectional view construction, and finally, the display options. The omnidirectional snapshots were shown, together with the hardware resource utilization of several image blending techniques.

It is shown that the processing demand is rather high even for modest panorama resolution and low resolution image sensors. The centralized approach for implementing real-time applications in multi-camera systems is not efficient in terms of processing. As an alternative, workload distribution and parallel implementations are encouraged for achieving high resolution and high frame rate reconstructions. Providing embedded workload distribution and parallelism capability in a multi-camera system requires innovation at the architectural level. In the next chapter a novel distributed approach is introduced for the realization of high-performance multi-camera systems using the very high-resolution cameras.

References

1. Anderson SF, Earle JG, Goldschmidt RE, Powers DM (1967) The IBM system/360 model 91: floating-point execution unit. IBM J Res Dev 11(1):34–53
2. Bouget J (2015) Camera calibration toolbox for MATLAB. http://www.vision.caltech.edu/bouguetj/calib_doc. Accessed 20 Oct 2015

3. Hartley RI, Zisserman A (2004) Multiple view geometry in computer vision, 2nd edn. Cambridge University Press, Cambridge
4. Meyer-Baese U (2007) Digital signal processing with field programmable gate arrays, 3rd edn. Springer, Berlin
5. Szeliski R, Uyttendaele M, Steedly D (2011) Fast Poisson blending using multi-splines. In: IEEE international conference on computational photography (ICCP). doi:10.1109/ICCPHOT.2011.5753119
6. Tierney J, Rader C, Gold B (1971) A digital frequency synthesizer. IEEE Trans Audio Electroacoust 19(1):48–57
7. Volder JE (1959) The CORDIC trigonometric computing technique. IRE Trans Electron Comput EC-8(3):330–334
8. Wilburn B, Joshi N, Vaish V, Talvala EV, Antunez E, Barth A, Adams A, Horowitz M, Levoy M (2005) High performance imaging using large camera arrays. ACM Trans Graph 24:765–776. doi:10.1145/1073204.1073259

Chapter 5
Miniaturization of Multi-Camera Systems

In this chapter, we present methods for creating and developing miniaturized high definition vision systems inspired by insect eyes. Our approach is based on modeling biological systems with off-the-shelf miniaturized cameras combined with digital circuit design for real-time image processing. We built a 5 mm radius hemispherical compound eye, imaging a 180° × 180° field of view while providing more than 1.1 megapixels (emulated ommatidias) as real-time video with an inter-ommatidial angle $\Delta\phi = 0.5°$ at 18 mm radial distance. We made an FPGA implementation of the image processing system which is capable of generating 25 fps video with 1080 × 1080 pixel resolution at a 120 MHz processing clock frequency. When compared to similar size insect eye mimicking systems in literature, the system described in this chapter features 1000× resolution increase. To the best of our knowledge, this is the first time that a compound eye with built-in illumination idea is reported. We are offering our miniaturized imaging system for endoscopic applications like colonoscopy or laparoscopic surgery where there is a need for large field of view high definition imagery. For that purpose we tested our system inside a human colon model. We also present the resulting images and videos from the human colon model in this chapter.

5.1 Introduction

Each year, over six hundred thousand people lose their lives due to the colon cancer in the world [32], which means more than 10,000 deaths per week in average. Even though the colonoscopy examination procedures are applied, there is a certain amount of miss rate in colonoscopy procedures due to narrow field of view (FOV) imagery employed in current systems. To illustrate the problem in colonoscopy, in Fig. 5.1 a simplified sketch is represented for the problem of narrow field of view in the colonoscopy applications. Since the human colon has a folded structure,

© Springer International Publishing AG 2017   89
V. Popovic et al., *Design and Implementation of Real-Time Multi-Sensor Vision Systems*, DOI 10.1007/978-3-319-59057-8_5

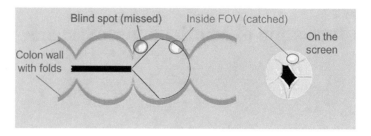

Fig. 5.1 The problem of narrow field of view in colonoscopy applications

Table 5.1 Current solutions for panoramic colonoscopy

Company	Solution	# cameras	FOV	Multi-camera image processing	Main drawbacks
Olympus [33]	Large FOV optics	1	N/A	No	Single camera limited resolution
Endochoice [12]	Multi-camera	3	330° × 330°	No	Separated windows hard to follow
Avantis [25]	Multi-camera	3	330° × 330°	No	Separated windows hard to follow
Naviaid [14]	Large FOV mirror	1	180° × 180°	No	Size doesn't allow working channel
Giview [10]	Large FOV mirror	1	180° × 180°	No	Size doesn't allow working channel

while the colonoscope moves towards the colon, it misses the behind-fold regions at the peripheral areas with respect to the forward and backward movement of the colonoscope.

Beyond the problem in colonoscopy, there are many works that show the need of large field of view, or panoramic imaging capability for endoscopy [8, 13, 19, 22, 24, 26, 34]. Most of them are optical systems with either mirror or wide angle lens-based solutions, which have main drawbacks as limited resolution, optical distortions, and lack of multi-perspective view. In [11], new technologies to overcome the narrow field of view problem in colonoscopy procedures are reviewed. In Table 5.1, a feature comparison of the current colonoscopy devices from different companies, Olympus, Avantis, Endochoice, Naviaid, and Giview, is given. For example, the solutions from Endochoice [12] and Avantis [25] utilize three cameras and do not have a compound panoramic image which ends up with three separated windows. Hence it results in difficulty for the operator to follow the whole imaging area. Other solutions from Naviaid [14] and Giview [10] utilize parabolic-like mirror solutions. Their monolithic optical solution occupies the whole diameter of the colonoscopy device due to their size and does not allow any working channels, which is required to remove the polyps or make small operations during examination. A recent solution from Olympus [33] accommodates a forward large angle of view lens and a surrounding parabolic mirror. The solution has a large FOV reflected onto a single image sensor, which makes it limited to the single image sensor's resolution. Therefore, the works and attempts from different companies show that there is a certain need for wide FOV imaging in colonoscopy domain.

The nature has its solutions for panoramic imaging. The compound eye structure of insects is a good example of this kind of solution, which has been a topic for biologists. In order to have a uniform and high resolution imaging capability in such applications, our proposal is to exploit the insect eye imaging to the endoscopy domain. For this purpose we propose a miniature multi-camera system, which also have the capability for 3D reconstruction for future applications like capsule endoscopy.

In the literature, there are two main approaches for implementing bio-inspired round shape compound eyes, which can be classified as curved optics and electronics for miniaturized systems [9, 15, 18, 30] and discrete component integration at macro level [1, 6, 7, 23]. The first group of methods suffer from diffraction limit and its effect on image definition, second method is not optimized and utilized for miniaturization due to non-optimal component sizes. The miniaturized multi-camera large field of view imaging systems have a wide variety of application areas like robotics, medical, and commercial products. The need for miniaturized bio-inspired vision systems for applications such as robotics and endoscopy is addressed in a recent review [27] where the current systems are presented comparatively.

In [30], a methodology is proposed for multi-lens curved imagers where authors described integration techniques of elastomeric compound optical elements with thin silicon photo-detectors and method for deforming these integrated sheets to get hemispherical surfaces. They reported a number of 180 bio-mimicked ommatidias, capable of imaging in a $360° \times 80°$ field of view and physical dimensions around 12 mm diameter. Authors compare their image definition capability with the eyes of fire ants (*Solenopsis fugax*) and bark beetles. They have also reported ray tracing simulation results but not real captured images with their proposed imaging system. Although this approach is promising for further miniaturization, it lacks the trade-off between the number of ommatidias and physical restrictions for building the photo-detector and the lens of each ommatidia. In[9], another micro-machining approach is proposed. In their system they have achieved $180° \times 60°$ with 630 ommatidias on a partial cylindrical shape and they reported inter-ommatidial angle of $\Delta\phi = 4.7°$ and acceptance angle of $\Delta\rho = 4.7°$. Their methods are based on complex integration and fabrication techniques. Both [30] and [9] are combination of micro-lens and micro-integration-based methods, which are trying to copy natural insect eyes one to one basis. Therefore, as their biological counterparts, they have the same problem of diffraction limitation, which ends up with low resolution with the given limited diameters [16, 17, 21]. Because of this problem, although they are promising for further miniaturization from the mechanical and manufacturing point of view, they have to tackle with the need of relatively high resolution despite the diffraction limit of the individual lenses.

In [1, 23] and [28], the methods proposed to manufacture multi-camera compound eye mimicking imaging systems; however, those systems are not optimized for miniaturization in imaging specific defined ranges. In the mentioned systems, the cameras are placed in a layered method without taking into account what is the desired distance of the objects that will be imaged and what is the physical sizes of the individual camera units.

In [2], a combination of mammalian vision and insect vision is proposed. It is a 3-camera array and the authors are focused on the lens design for individual image sensors and the image processing is done offline by using free stitching software. They reported a maximum FOV of 130° and final pixel resolution of 643 × 366.

For endoscopic applications the illumination capability is an obvious requirement. Beyond the limitations of previous insect eye inspired systems, there is no illumination method proposed before for insect eye inspired imaging systems to work in dark environment.

We propose a method to build miniaturized high definition bio-inspired vision system, which eliminates the deficiencies of previous approaches in terms of image resolution and physical size. Moreover, we propose a distributed illumination method, which is beyond the capabilities of natural compound eyes [17], giving our imaging system the capability of working in dark environments for proximity imaging applications such as endoscopy. We built a 5 mm radius of curvature hemispherical compound eye, imaging in a 360° × 90° or 180° × 180° field of view while providing nearly a total number of 1.1 million emulated ommatidias with an inter-ommatidial angle $\Delta\phi = 0.5°$ at 18 mm radial distance. Our FPGA implementation of the system is capable of generating 25 fps video with 1080×1080 pixel resolution. We also compare our system with the dragonflies [29] not only in terms of image definition but also in terms of size. To the best of our knowledge, this is the highest reported definition capability with real-time video output at this scale of miniaturization for insect eye mimicking camera systems and this is the first time that a compound eye with illumination idea is reported.

Our approach described here includes analysis of previous multi-camera component integration methods from miniaturization perspective, definition of a model for image generation that is inspired from the insect vision, and FPGA implementation of the desired image processing blocks as digital circuits which can be also miniaturized by a future ASIC design.

5.2 Opto-Mechanical System Design

5.2.1 Effect of Single Camera Dimensions

The dimensions and volume of each individual camera unit play an important role in the miniaturized level multi-camera system. Therefore, the primary criteria for choosing the individual cameras that will be used in the final construction are its volume and dimensions. In [1], the single camera units are modeled as circular shapes that will be placed on a hemispherical surface. Thus, the third dimension of the cameras is omitted since the volume of the individual cameras V_{cam} is negligible when compared to the hemispherical structure volume, $V_{hemisphere}$. For the miniaturized model, the volume and shape of the cameras are determining factors. Hence, the selected individual cameras are getting more importance. The minimality and simplicity of the individual cameras are also important in terms of having a minimal processing circuit. The electrical specifications of the cameras for embedded system and image processing circuit design are considered in Sect. 5.3.1. The limits for the dimensions of the image sensor array are mainly dependent on the pitch size of an individual pixel. On the optics side, a single minimal lens design and an appropriate pinhole (aperture) for desired resolving capability at a desired distance are required. The optical and mechanical properties of the chosen individual camera are given in Table 5.2. The mechanical illustration of the individual camera unit is given in Fig. 5.2a.

The vision capability of the camera system is bounded by the vision capability of each individual camera unit. The chosen cameras are utilizing a pinhole camera with boro-glass optic lens, where the pinhole aperture diameter is $d_{pinhole} = 110\,\mu m$ and a focal length of $f_{pinhole} = 660\,\mu m$. According to the Rayleigh criterion, the theoretical capability of separating two point source airy discs at image plane is given by: $x_i = 1.22 \times \lambda \times \frac{f_{pinhole}}{d_{pinhole}}$. For the visible spectrum, a minimum $\lambda_{min} = 420\,nm$ can be assumed. Hence an $x_i \approx 3\,\mu m$ is obtained. The CMOS image sensor of the chosen camera is 250×250 photo-sensor array with a sensor pitch size $3\,\mu m$. Therefore, the optical capability is matching with the electronic part of each individual camera unit. If we assume the two point sources are separated by an angle of θ_0, this angle can be interpreted as $\theta_0 = \frac{x_i}{f_{pinhole}}$. As a result, the angular separation

Table 5.2 Single image sensor opto-mechanical specifications

Specification	Value
Size	1 mm×1 mm×1.85 mm
F# number	6.0
FOV (vertical× horizontal)	64° × 64°
Focal length	0.66 mm
Aperture	0.11 mm
Depth of focus	3–50 mm
Pixel array	250 × 250

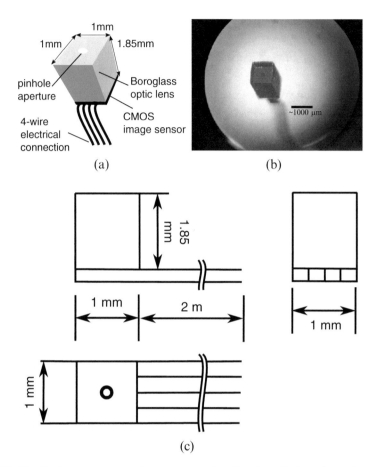

Fig. 5.2 The single camera chosen for implementation (**a**) physical dimensions, (**b**) close-up photo, (**c**) 3-view dimensions of the camera

capability of the individual cameras is calculated as $\Delta\phi = \theta_0 = \frac{3\,\mu m}{660\,\mu m} \approx 0.26°$ at a depth of field of 3–50 mm, which is given as the best focus distance of the individual cameras that are utilized for the system.

Since the resolving capability will determine the limit of the resolving capability of the whole system, an initial comparison with the existing miniaturized camera systems and natural counter parts leads to the choice of the individual cameras in terms of resolution. As a comparison, the system implemented using micro-machining techniques and special materials in [30] has a resolving capability of $\Delta\phi = 11°$. For the system in [9], the resolving capability is reported as $\Delta\phi = 4.7°$ where for a natural system, which is known as a superior resolution insect eye, the dragonfly eye has a resolving capability of $\Delta\phi = 0.24°$ [29]. Therefore, the resolving capability of the chosen system is reasonable when compared to the state-of-the-art systems and natural systems. Of course, the size and quality of the used optical design have an effect on the optimal focus distance of the camera. For the camera that is chosen for implementation, the optimal focus distance is 3–50 mm, which is an acceptable distance for applications like endoscopy where the objects of interest are around 5–50 mm.

As a conclusion, the minimal dimensions and reasonable resolving capability of the individual camera unit chosen for the system implementation are determined as appropriate and it is used in the rest of the design.

5.2.2 Proposed Camera Placement for Miniaturized Camera Model

The first target for our model is to have many cameras placed on a curved surface and looking outside of the center of the curved surface. A hemispherical surface is chosen, which is suitable with natural counterparts. Then the question that arises is how to place the cameras optimally on the surface. In [1], there is a layer by layer placement of the cameras proposed, which is not optimized for miniaturization.

There can be different approaches to place the cameras on a dome according to the application needs. One approach is to minimize the number of cameras while having an overlap at a certain distance determined by the application from the camera surfaces or hemispherical compound camera center. By this approach, total cost, power consumption, and the total dimensions of the final hemispherical camera can be minimized. The other approach can be maximizing the number of cameras that can fit into the given total dimensions of the hemispherical volume. In the later approach, there will be more overlap of the camera angle of views, hence closer observable distance and better 3D infrastructure.

To start the placement, the camera positions can be defined as circles on a hemisphere as illustrated in Fig. 5.3a. In this way, the interference of the adjacent cameras' base area is prohibited even at the corners of the two cameras' square base area. Moreover, there will be a room left for the illumination channels. However, unlike the arrangement in [1], the hemispherical surface we choose is not the final outer surface of the hemispherical structure. Instead, it will be an inner surface with a radius $r_{hin} = r_{hout} - l_{cam}$. Here the l_{cam} is the length of the camera from its optical pinhole aperture to the end of the CMOS image sensor which is the omitted dimension in the previous work [1]. This new model is illustrated in Fig. 5.3a.

Then the cameras are first placed on a quarter of the inner hemisphere without having interference with the circular surfaces of the cameras. The initial placement will be in one dimension to cover 90° on one quarter of the semi-circle on the y–z plane as shown in Fig. 5.3b. Here, there might be two choices as mentioned before: minimize the number of cameras or minimize the distance of overlap of the camera AOVs by maximizing the number of cameras. For the final design, maximizing the number of cameras is chosen for two reasons. First, to fit in a limited dimension dictated by the application and to have a reasonable AOV overlap

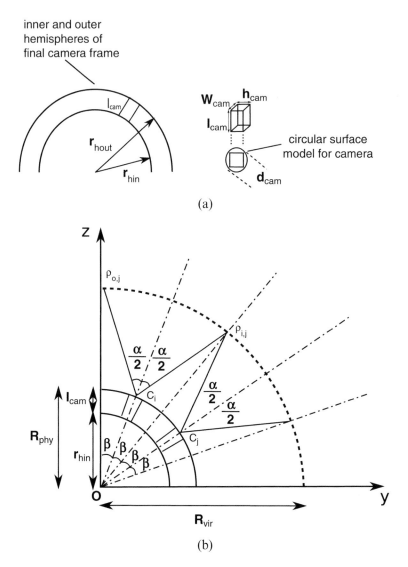

Fig. 5.3 New camera placement model (**a**) single camera circular surface model on dome y–z plane. (**b**) Geometrical relations of the cameras in one quarter of the dome in y–z plane

distance suitable for endoscopic applications. Second, to explore the limits of our off-the-shelf component-based approach and resolution capability in the millimeter scale devices domain. In this way, the proposed system can be a reference for the further miniaturization targets which are desired to be achieved by using material and micro-machining techniques.

As an initial application, the colonoscopy imaging is chosen. The human colon is about 40–60 mm in diameter. Then the acceptable object distances will be at

20–30 mm range from the hemispherical camera center. The acceptable image device diameter will be at 10 mm range so a $R_{phy} = 10$ mm is chosen, which will be the final outer radius of our hemispherical structure. Also the length of the camera l_{cam} is assumed as 2 mm instead of its actual dimension 1.85 mm, for manufacturing margins. For the same reason, the side dimensions for the square hole openings where the cameras will be inserted are taken as 1.1 mm.

In order to maximize the number of cameras, the angle between each camera which is given as 2β in Fig. 5.3b should be minimized. Concurrently, the whole 90° FOV between the y–z axes should be covered at least one of the cameras angle of view, α. The relation between β and α can be obtained by the sine law on the $\widehat{oc_i\rho_{i,j}}$ triangle. This relation is given by (5.3)–(5.4). Here, the R_{vir} is the desired radial distance from the hemisphere center o to the point where adjacent camera angle of views intersect. R_{phy} is the physical constraint for the actual radius of the hemisphere, which is also given as the $r_{hout} = r_{hin} + l_{cam}$ as shown in Fig. 5.3a. The constraint is to fit as many cameras as possible on the quarter circle on the y–z plane of hemisphere with inner radius r_{hin}. The diameter of the cameras' base area of the circumscribed circle shown in Fig. 5.3a, given by (5.1). Then the number of cameras in one quarter in one dimension on the y–z plane is obtained by dividing the arc length to the d_{cam} as given in (5.2). Since this division gives the maximum number of cameras that can be fit on this arc and not necessarily an integer number, we take the flooring function to this number.

$$d_{cam} = \sqrt{w_{cam}^2 + h_{cam}^2} \tag{5.1}$$

$$N_{camv} = \left\lfloor \frac{\frac{\pi}{2} r_{hin}}{d_{cam}} \right\rfloor \tag{5.2}$$

$$\frac{\sin(\pi - \alpha/2)}{R_{vir}} = \frac{\sin(\alpha/2 - \beta)}{R_{phy}} \tag{5.3}$$

$$\beta = \alpha/2 - \sin^{-1}\left[\frac{R_{phy}}{R_{vir}} \sin\frac{\alpha}{2}\right] \tag{5.4}$$

Then we extend this approach by defining longitudinal circular layers as illustrated in Fig. 5.4 passing through the camera locations on the quarter hemisphere. Here, the radius of each lateral circle is geometrically related to the β angle given by (5.5).

$$r_i = r_{hin} \sin\left[(2i - 1)\beta\right], \quad i \in [1, N_{camv}] \tag{5.5}$$

$$N_{camhi} = \frac{2\pi r_{hin} \sin\left[(2i - 1)\beta\right]}{d_{cam}} \tag{5.6}$$

In this way, the maximum number of cameras in horizontal (N_{camh}) and vertical (N_{camv}) directions is determined. For the cameras described above, the equations

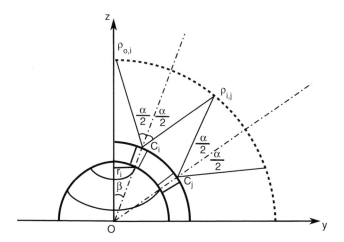

Fig. 5.4 Illustration for the layers for extending the camera placement around the hemisphere

through (5.1)–(5.6) result in $N_{camv} = 3$ and $N_{camh1} = 3$, $N_{camh2} = 9$ and $N_{camh3} = 12$, equal to 24 cameras in total. The longitudinal β angles of the cameras in one quarter are obtained as $\beta = 15°$. The analysis on the minimum distance required for overlapping field of views of the neighboring cameras is done according to the method from [1] and resulted in a $R_{vir} = 18$ mm. To achieve the same overlap distance, the method for camera placement in [1] results in 29 camera positions and a 6 mm radius hemisphere.

Then we use a commercial mechanical modeling tool to generate the actual final structure, which is shown in Fig. 5.5a. The mechanical model is then fabricated by using a 5-axis CNC machine. The final fabricated prototype is shown in Fig. 5.5b.

5.2.3 Calibration

For the calibration process, we use a SIFT[20]-based image bundle adjustment commercial software. The SIFT-based software utilizes the key point extraction and feature matching and pinhole camera model to calculate intrinsic and extrinsic parameters of each camera. Then we use a MATLAB script to calculate the camera coordinate vectors and pack the parameters as look up table to feed the hardware system. The calibration is done offline and the parameters are used from these look up tables during online processing.

To do the calibration, we first take shots inside a 35 mm diameter cylindrical closed tube with texture. After collecting the images from 24 cameras, a construction is performed in the software environment to determine the extrinsic parameters which are defined with yaw (α), pitch (β), roll (γ) and intrinsic parameters,

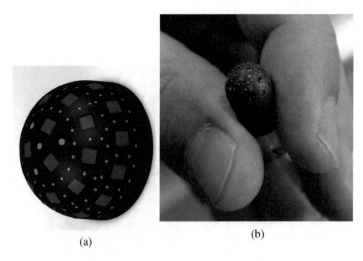

(a)

(b)

Fig. 5.5 The camera placement and the mechanical design for the proposed miniaturized multi-camera endoscopic compound eye system: (**a**) The fabrication ready positions of the cameras (shown in *red*). (**b**) Fabricated prototype

focal length (f_L), principle point (cc), k parameters (k_1, k_2, k_3) for lens distortion correction. Then from yaw, pitch, roll rotations, we define a camera coordinate system with three unit vectors, $(\mathbf{t}, \mathbf{u}, \mathbf{v})$ as shown in Fig. 3.4a.

The diameter of the calibration tube is important and should be compatible with the final application since the calibration process can give slightly different results due to the geometrical relation of the individual cameras and the optimal focus depth of each camera. For example, we choose a 35 mm diameter tube which is compatible with the human bowel inner diameter.

5.2.4 Neural Superposition Virtual Ommatidia

The neural superposition type eyes of insects are combining the intensity information from different optical channels and form an erect single image [17]. This is the base for the second part of the modeling of the insect eye concept for the method described in this chapter. After the camera placement and calibration is done, for generating the final composite image, each direction in the surrounding space is defined as a virtual ommatidia (VO). Each VO is supposed to sample a light ray from space to represent the intensity value at that ray direction. Then the whole panoramic image will be the combination of the intensity values of these VOs. In this way, the problem of generation of the intensity values of each virtual ommatidia is reduced to a light ray tracing problem. Then for each VO, during the real-time operation, optical channels (cameras) that are contributing to the direction of that VO should

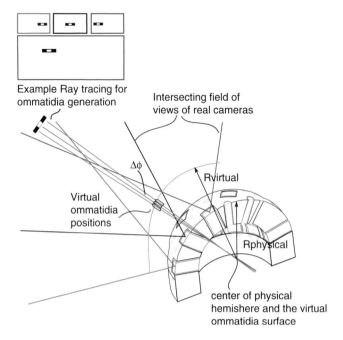

Fig. 5.6 Virtual ommatidia sampling concept with prototype

be selected. Then a superposition of the candidate intensity values from the selected cameras should be done. Therefore, on the image processing side, to generate the VO intensity values, we can utilize the vector-based ray tracing method described in [1]. Here, we use the ray tracing method on one half of a unit sphere, which is then assumed to be positioned at a relatively far distance from our compound eye, $R_{virtual}$, as shown in Fig. 5.6. On the hemisphere with radius $R_{virtual}$, the virtual ommatidias are formed without any spaces between them. So the inter-ommatidial angle $\Delta\phi$ can be obtained by dividing the number of ommatidias on the perimeter one direction of the virtual hemisphere of the compound eye by 180°.

For ray tracing to generate a final intensity value per virtual ommatidia, first there is a need to determine the two angles θ and φ on a spherical coordinate system, which has its origin at the center of the insect eye camera. For that, the angles are generated starting from 0 and scan the whole hemisphere with a certain number of steps, which corresponds to the $\Delta\phi$ in the insect eye model. After this step, we have to find which cameras are seeing this direction. For this purpose, a dot product of the direction vector $\boldsymbol{\omega}$ and each camera unit vector $\mathbf{t}$, which represents the viewing direction of the camera, is performed. If the angle between the camera and the ommatidia vector $\boldsymbol{\omega}_j$ is less than half of the angle of view of the ith camera, this implies that the camera i contributes to the intensity value of ommatidia represented by $\boldsymbol{\omega}_j$. After this step, an estimation of the ray position on the each contributing camera image frame is achieved by projecting $\boldsymbol{\omega}$ onto the camera coordinate system determined by **t**, **u**, and **v**. For the superposition step, methods can be found in [1, 7, 23]. The method described previously in [7] is utilized in this work, which gives better parallax and seam performance.

5.2.5 Illumination

No matter which of the methods in [1, 9, 30] are taken into account, there is not an illumination method proposed before for compound eye systems to be able to operate in dark conditions especially for close distance imaging such as endoscopic imaging. This feature, which is unique to our compound eye model, does not exist even in the natural compound eyes [17]. As seen in Fig. 5.5a, b, we utilize the empty spaces around the cameras on the hemispherical frame to place the cylindrical holes for fiber optic illumination channels. For this purpose we use plastic optical fiber (POF), due to its flexibility and low cost. We use fiber optic illumination channels reaching from a light source 2 m away to the hemispherical camera tip. It gives an opportunity to illuminate the targets at a proximity to the hemisphere. This feature can be utilized for the applications like endoscopy or any other dark environment applications for robotics. In our current prototype we have 112 illumination channels distributed around the cameras to the empty spaces on the hemispherical frame. The theoretical model for illumination capability can be given by (5.7) [17].

$$F = \frac{L \times A_e \times A_r}{d^2} \text{ lumen} \tag{5.7}$$

In (5.7), we define the area as the surrounding surface of the bowels. d will be approximately 20 mm from the surface of our imaging system and 25 mm from its center. If we assume a 25 mm radius hemispherical area around our imaging system, the total area will be $A_r = 2 \times \pi \times 625$ mm^2, which is 1250 mm^2.

L is the luminous flux of the light source. A_e is the total area of the emmiter. We use a commercial light source with 6000 K white LED illumination, which is capable of delivering 700 lumen for a diameter of 13.5 mm, equivalent to $A_e = 143.1$ mm^2. Our actual emmiter area is the total area of the fiber opening of the illumination system. We have 108 channels with 250 μm diameter and 4 channels at the top with 500 μm diameter. So the total area of our emmiter channels is $A_c = 6.1$ mm^2. Then the actual L for us will be proportional to the area ratio between the light source and our total channel area multiplied by the luminous flux of the light source: $\frac{A_c}{A_e} \times 700 = 29.8$ lumen. By substituting L with this number in (5.7), we end up with the light delivering capability of our system at 20 mm from the surface of the imaging system on a hemispherical area $F = 93.2$ lumen.

5.3   Circuit and Embedded System Design

The whole prototyping chain with the used tools is summarized in Fig. 5.7. The built system as a whole with the 24 cameras hemispherical tip, cabling, Xilinx FPGA evaluation board, and the in-house designed custom interface PCB is shown in Fig. 5.8. The human colon model used during the experiments is also shown next to the system in Fig. 5.8. In the following subsections, the embedded system and the hardware components are described which reside inside the embedded system.

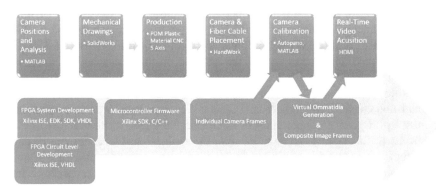

Fig. 5.7 Whole prototyping chain for the system

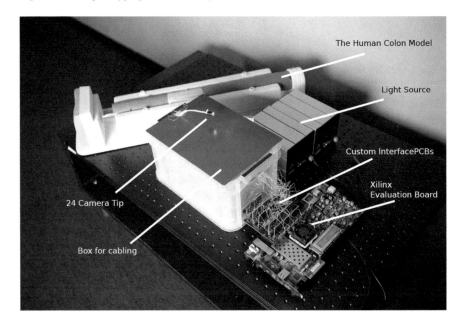

Fig. 5.8 The complete system used for experiments

Table 5.3 Single image sensor electrical specifications

Specification	Value
Shutter mode	Rolling
ADC resolution	10-Bit
Data interface	LVDS signal, 10 bit signal + 1start 1stop bit
Operation mode	Free running
Frame rate	44–56 fps
Pixel array	250 × 250

5.3.1   Single Camera Interface

The summary of the specifications for the selected single camera is given in Table 5.3. We design the camera interface as a custom IP compatible with Xilinx AXI peripheral bus. The downstream image frame data comes as a serial data stream via an LVDS pair with 200 Ω line impedance, which is not standard LVDS.

The frame rate and exposure time are controlled by altering the supply voltage of the internal PLL of the camera between 1.8 and 2.4 V. We designed and built a PCB for interfacing each camera. The PCB contains a fast comparator to sample and convert the LVDS line to 1.8 V 100 Ω standard LVDS, which we can connect directly to the LVDS inputs of FPGA. On the PCB, we also place an adjustable voltage controller with a digitally controlled resistor at its adjust node. In this way, the supply voltage of each camera can be controlled digitally by the FPGA. The camera PCB units are connected to the FPGA through another custom PCB which has a FPGA Mezzanine Card (FMC)connector compatible with the FPGA board we used.

Inside the FPGA, our single camera interface samples the incoming serial data with a 200 MHz clock and decodes the serial stream by detecting the edges and pulse widths. The encoding used in the serial stream is a derivative of Manchester encoding. The average frequency of the camera signal is around 30 MHz. After sampling, we deserialize the pixel stream and get 10-bit precision Bayer filtered pixel intensity values. Then we feed the pixel values into two different paths. One of them is frame memory which is accessed by the virtual ommatidia generation unit. The second path is converting the video stream to Xilinx AXI-stream and feed the data to DDR3 SDRAM interface through Xilinx video DMA units. This latter path is just to get the single camera streams visually through HDMI interface for making the calibration step and showing the single camera videos individually as well during the operation. Examples for single capture from the system are shown in Fig. 5.11 during operation.

5.3.2 System Level Design Considerations

The system is targeted for Xilinx VC707 board. This board has an FMC connector
that can be used for camera connections, RS232 serial port for user interface,
and HDMI output for video stream output. As the next step, we target an ASIC
design that can be packaged with the miniaturized hemispherical head, so we tried
to minimize the external components needed for the system. To that end, we use
on-chip block RAMs as the camera frame memories which are accessed by the
panorama pixel generation block unlike the implementations in [1, 7, 23]. For
streaming out the single camera images and the panoramic image through HDMI,
we used the DDR3 SDRAM available on the board. However, for our ASIC design,
we are aiming to use a simple parallel video interface similar to any conventional
image sensor available on the market. The block diagram for the designed system is
shown in Fig. 5.9.

5.3.3 Image Processing Hardware

There are different hardware approaches in literature for ray tracing-based panorama
generation methods [1, 23, 28]. In [1] and [23], a central processing unit generates

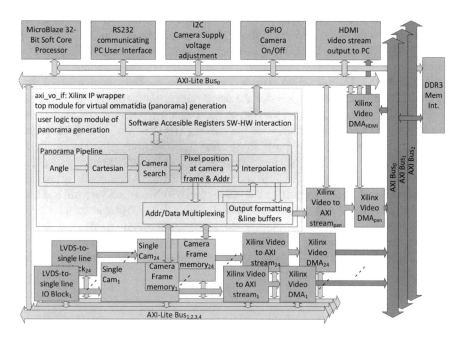

Fig. 5.9 FPGA System simplified block diagram, the custom designed modules are shown in
green, where the Xilinx IP and busses are shown in *blue*

the final panorama by utilizing a pipeline structure. In [28], a distributed approach
is reported, where each node has a camera connection and sends the pixel values to
a central unit via a network on chip. Since we are targeting a miniaturized system
which should be also affordable in ASIC design, we also utilize a pipeline structure.
In this approach, the number of cycles spent for generation of one output pixel of the
panorama has a significant impact. In the end, the total number of cycles is equal to
$\text{cycles}_{\text{pix}} \times w_p \times h_p$. The $\text{cycles}_{\text{pix}}$ is the number of cycles spent for generation of one
panorama pixel; w_p is the width; and h_p is the height of the final panoramic image
in terms of number of pixels.

For the pipeline design, a previous implementation methodology described in [1]
is utilized. For each panoramic image, the pipeline is fed with the corresponding
latitude and longitude angle, starting from the top left corner of the image. Then it
calculates the Cartesian vector for that direction. Later it finds the candidate cameras
looking in that direction by searching in all the cameras in the system where each
camera check brings 1 clock cycle. This brings an unnecessary number of clock
cycles since there is an upper limit of the number of cameras looking at a certain
direction in space due to the spherical placement of the cameras. However, the
maximum number of cameras was bounded to 20 in the previous approach [1]. So
the design of the pipeline for the 24 camera system is performed by retiming. Then
the system is mapped onto the Virtex-7 FPGA device.

The pipeline structure is described in Fig. 5.10. The throughput limit of each main
block is shown as minimum required clock cycles which are needed to complete the
operation and get ready for the next one. In Fig. 5.10, each color is representing a
different main block in the pipeline. For example, the red colored unit is the angle
generation unit and the minimum clock cycles required for that unit is $T_h = 2$ cycles.
The throughput or the minimum clock cycles required to generate the intensity
value of one pixel in the final panorama T_{hsys} are dictated by the largest value in
the pipeline $\max \{T_h\}$.

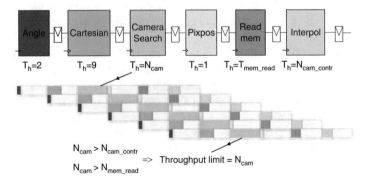

Fig. 5.10 The panorama generation pipeline analysis of the system

Then assuming a 25 fps for real-time video requirement, for an $N \times M$ final panorama image the real-time operation clock frequency is defined as $F_{min} = 25 \times N \times M \times T_{hsys} \times 10^{-6}$ MHz. For example, for a system with 20 cameras, to generate a 1024×1024 pixel panoramic video at 25 fps, a minimum of 525 MHz operation clock is required.

The first implementation in the scope of book is done with the constraints above with a 100 MHz clock frequency. Then the obtained maximum frame size at 25 fps is observed as 166 Kpix, 408×408 video for $180° \times 180°$ representation and 730×182 video for $360° \times 90°$ representation.

The designed system block diagram is given in Fig. 5.9. In total, there are 5 AXI-lite buses in the system. The 24 camera interfaces are utilizing 4 AXI-lite buses and each of these AXI-lite buses is connected to 6 camera interfaces. Each AXI-lite bus has 32 bit data width.

We applied a parallelism for camera search by utilizing a loop unrolling technique while checking the number of cameras seeing the particular direction. Instead of making the dot product operation in (3.5) for each camera in a loop of 24 cameras, we utilize 24 parallel dot product operations and 24-to-5 encoder to determine all the contributing cameras in n_c clock cycles. The n_c is the number of contributing cameras to a particular direction. For the system propose, there are 24 cameras and we made an analysis of the system to determine the maximum number of cameras which can contribute to any particular direction. When we have done the analysis, it is figured out that at any direction, at most four cameras can contribute to generation of the intensity value for that direction. Therefore, by this low level parallelism, a $6\times$ frame rate increase is achieved with a given clock frequency when compared to the systems proposed in[1, 23].

All the sub-blocks concerning the single camera interface and panorama generation unit are written in VHDL. The method for development is to first simulate the VHDL entities in Modelsim simulation environment and then use Xilinx AXI slave interface at Xilinx EDK environment to embed the sub-blocks in the system. The full system is realized in the Xilinx EDK. For the implementation of the standard IPs such as Xilinx VDMAs and bus components, we used the EDK tool of Xilinx.

An evaluation for migrating the design to an ASIC prototype is done by using tsmc40 nm technology with the minimal requirements without using the microprocessor and bus components. The single camera interfaces, panorama generation unit, and a simple controller in the form of an FSM are utilized in this minimized design. The initial implementation for feasibility of the design resulted in a die area of 10 mm $\times$ 7 mm and 49 million 2-input NAND equivalent gate count. The power analysis and fabrication of the chip are targeted for future work.

5.4 Results

5.4.1 Visual Results

In Fig. 5.11, an example output from the realistic colon model is shown in 1024 × 1024 resolution. The single 24 camera streams and the compound panoramic image can be seen together in this figure. As seen on the 180° × 180° compound panoramic image, the polyps at the sides, even behind the folds of the colons can be easily captured by our system. In Fig. 5.11, the red box shows a 140° × 140° FOV, which is the FOV that can be provided by most of the colonoscopy devices in the market today. Two video output streams of the system are provided on the supplementary material. On the supplementary videos, the resolution of the panoramic video streams is 1080 × 1080 pixels resolution with 190° × 190° field of view to show the boundaries of the cameras. For the initial prototype, 30 illumination channels at the top part of the system are assembled manually. The rest of the fiber channels at the lower part of the hemisphere are attached on a ring of 14 mm diameter and attached to the outer shell of the 10 mm imaging tip. The number and viewing angles of the fiber channels on the experimental illumination ring are same as the number of channels on the actual tip. Therefore, the illumination capability of the fiber channels on the hemispherical tip is preserved.

5.4.2 Efficiency of the System Size

In [16], it is shown that for diffraction limited compound eyes, the relation of the eye radius, the acuity parameter $\Delta\phi$ and the light wavelength λ, is given by (5.8). The mapping between the insect eyes and human eyes is shown in Fig. 5.12. Here,

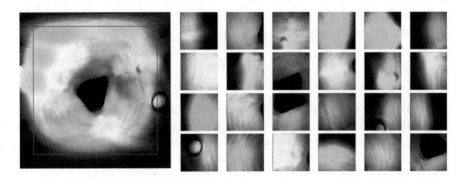

Fig. 5.11 An example of the output image from the system, the panoramic image is 180° × 180° FOV, 1024 × 1024 pixel resolution, each single camera images is in 248 × 248 pixel resolution with 64° × 64° FOV. The *red box* on the panoramic image indicates a 140° × 140° central portion of the image

Fig. 5.12 The relation
between insect and human
eye resolving capability

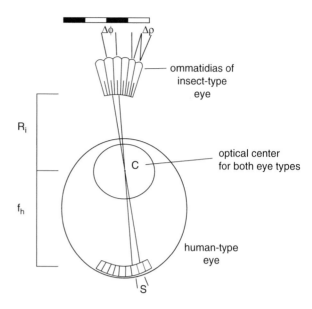

the $\Delta\phi$ is the acuity or resolving power of the both visual systems and $\Delta\rho$ is the
angle of view of each single imaging unit.

$$R = \frac{\lambda}{2\Delta\phi^2} \tag{5.8}$$

By using (5.8), it is possible to estimate the expected ideal radius of curvature of
a compound eye by using the sampled wavelength and the given inter-ommatidial
angle of the compound eye. For example, given inter-ommatidial angle $\Delta\phi$ of the
systems in [30] and [9], the expected radius of the defined compound eyes should
be 0.06 mm and 0.046 mm, respectively. Both systems are realized as 6 mm radius
nearly. From this point, we can define a size efficiency parameter for the design and
implementation of compound eyes as (5.9).

$$\zeta_c = \frac{R_{imp}}{R_{ideal}} \tag{5.9}$$

In (5.9), the R_{imp} is the empirically measured or fabricated radius and the R_{ideal}
is the estimated value by using (5.8). Our system is implemented at 5 mm radius for
our 24 camera prototype. We have measured the acuity by using USAF 1951 line
pair pattern as 0.0087 rad (0.5°). An example image taken with USAF line pattern
is shown in Fig. 5.13. As (5.8) indicates, we have an expected radius of 3.3 mm. In
other words, if such a system exists in nature, it should be ideally sized around this
radius of curvature for the compound eye. Finally, our system has a physical radius
of 5 mm. However, there is a need for certain distance from the real cameras so that
their angle of views can intersect and cover the whole hemispherical field of view.

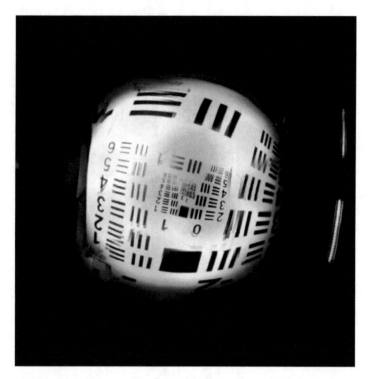

Fig. 5.13 An example image taken with USAF 1951 line pattern at 18 mm distance from the hemispherical compound eye center

We defined this virtual radius by our camera placement as 18 mm, where the virtual ommatidias can appear without any space between them. So we have an efficiency of 18% for our implementation.

5.4.3 Comparison with Different Insect Eye-Based Systems

In Table 5.4, we compare our system with the curved lens array-based compound eye systems and with the natural compound eyes. In this comparison, the main criteria are size, field of view, and resolution. Table 5.4 shows that our system features a significant resolution increase of 1000× when compared to the systems which have even larger radius of curvature. As seen from Table 5.4, our system is closer to natural systems in terms of resolution/size efficiency, which is very important for miniaturized wide FOV imaging especially for applications like colonoscopy [8].

Table 5.4 Comparison with insect eye systems in terms of resolution and size

Approach	Size (radius of curvature) (mm)	Number of ommatidia (pixels)	Field of view	$\Delta\phi$	$\Delta\rho$	R_{ideal} for 500 nm wavelength
Dragonfly [29]	5–6	30,000	180° × 160°	0.24° (0.0042 rad)		5–6 mm efficiency: 100%
[30]	6–7	180 (12 × 15)	360° × 80°	11° (0.19 rad)	9.7°	0.06 mm efficiency: 1%
[9]	6.4	630 (42 × 15)	180° × 60°	4.2° (0.073 rad)	4.2°	0.046 mm efficiency: 0.7%
This work	5	1920 × 512 1080 × 1080 ($\approx$1,100,000)	360° × 90° 180° × 180°	0.5° (0.0087 rad)	0.5°	3.3 mm efficiency: 18%

Table 5.5 Virtex-7 FPGA
resource usage

Resource type	Total count	Utilization ratio (%)
Virtex-7 slice LUTs	150K	50
Virtex-7 slice registers	154K	25
Total occupied slices	60K	80
BRAMs	31Mbit	94
DSP48s	185	6

5.4.4   Hardware Implementation Results

In our hardware implementation we used the VC707 evaluation board from Xilinx, which has the FPGA device XC7VX485T-2FFG1761C. The implementation results are given in Table 5.5 in terms of resource usage. Our next target is to implement the design as an ASIC. Hence, we are using on-chip memory resources for frame buffering. Therefore, the highest utilization is on the block RAM resources.

The design has different clock domains as 200, 100, and 400 MHz. The 400 MHz clock is for DDR3 RAM interfacing, 100 MHz is for Microblaze and AXI bus system, and 120 MHz for panorama video generation block. 200 MHz is the input sampling frequency of the camera interfaces. The output of the panorama generation unit is also RGB565 and it is converted to 24-bit by zero padding. The output of the system is 24-bit RGB pixel video stream. The highest video throughput from the panorama generation block is 465 Mbit/s, at 25 fps, 120 MHz, 1080 × 1080 spatial resolution. However, due to the AXI-Bus bandwidth limitations, the HDMI video output stream is limited to 24 fps.

5.5   Discussion and Future Directions

With the designed miniaturized hemispherical imaging system, we explore the implementation limit of the miniaturized compound eye mimicking systems. In this way it is shown that by using current off-the-shelf components and simple fabrication techniques, it is possible to achieve high definition and very large field of view compound eye system. The system proposed is comparable to the natural counterparts to previously proposed advanced material and micro-manufacturing-based systems with superior resolution capability. A previously proposed camera placement approach from literature is optimized for miniaturization. Moreover, the idea of a distributed, built-in illumination feature is provided and tested for compound eyes for better vision in dark environments. The further applications for the implemented design beyond endoscopy can be smart phone industry where a panoramic miniaturized imager is needed. The designed HW system allows changing the resolution and number of virtual ommatidias in real-time dynamically, hence, another application domain for the presented system can be to model and characterize different insect species' vision systems. So with this small size and the image processing capabilities, the proposed solution is also offering a ready to use framework to help insect eye research works in future.

The proposed calibration method is dependent on the texture of the scene that is captured by each of the cameras. This means the pixel resolution of each individual camera has an effect on the quality of the calibration process, i.e., determining the relative positions of the cameras to each other and the intrinsic parameters of each camera. We could also rely on the physical positioning of the cameras, however, as mentioned, then this would be misleading due to the fabrication errors. As the resolution of the individual cameras decreases, the quality of the calibration decreases proportionally. As a limit, 1 pixel cameras could be used for each individual camera, which is equivalent to the approaches in [9, 30]. Then, there is no possibility to make such a calibration, based on texture captured by each individual camera unit. Therefore, the methods like [30] and [9], which utilize very low (usually 1) pixel resolution for individual cameras, need very precise mechanical placement or special fabrication of each individual unit. On the other hand, for systems like proposed in this work, there is no need for high precision in manufacturing since there is the option to be calibrated subsequently, thanks to the high pixel resolution of each individual camera unit.

Therefore the precision of the placement is affected by this hand placement and fixing method. However, since we use a post-calibration method for defining the relative positions of the cameras and the intrinsic optical characteristics of each individual camera such as focal length, we overcome errors due to imprecise manufacturing at image processing level.

Our future direction is to focus on automatic polyp detection. In literature, there are different ways to detect polyps automatically which relies on single camera images [3–5, 31]. Since there are no such devices yet established as the system proposed in this work, no methods exists to use multi-view geometry features for detecting polyps during real-time video examination.

Another future path is to implement the electronic system as an ASIC design with necessary modifications in order to have our imaging system as a fully compact system. By this way the device that we fabricated can be applied to stand-alone applications like capsule endoscopy or robotics.

5.6   Conclusion

In this chapter, we present a novel insect eye system, which is comparable to its natural counterparts in size and resolution. The system also has the new feature of distributed illumination capability, which is even not found in biological systems. With its size, illumination, and resolution capabilities, our system can be an alternative camera for the endoscopic applications where there is a need for large field of view and high resolution. With our video reconstruction method inspired from the neural superposition type insect eyes, the whole area of interest can be visualized in a compact representation for ease of human observers who use the device for endoscopic examinations. When compared to its microlens-based counterparts, our system offers smaller radius and 1000× more spatial resolution. From a size/resolution efficiency point of view, our system is the closest model to its biological counterparts when compared to the current systems in literature. The system level design of the electronic parts and their implementation results are presented, as well as the visual results in a realistic human colon model where polyps at the side regions and even behind folds can easily captured.

References

1. Afshari H, Akin A, Popovic V, Schmid A, Leblebici Y (2012) Real-time FPGA implementation of linear blending vision reconstruction algorithm using a spherical light field camera. In: IEEE workshop on signal processing systems, pp 49–54. doi:10.1109/SiPS.2012.49
2. Aldalali B et al (2013) Flexible miniaturized camera array inspired by natural visual systems. J Microelectromech Syst 22(6):1254–1256
3. Ameling S, Wirth S, Paulus D, Lacey G, Vilarino F (2009) Texture-based polyp detection in colonoscopy. In: Bildverarbeitung für die Medizin 2009. Springer, Berlin, pp 346–350
4. Bernal J, Sánchez J, Vilarino F (2012) Towards automatic polyp detection with a polyp appearance model. Pattern Recogn 45(9):3166–3182
5. Bernal J, Sánchez J, Vilarino F (2013) Impact of image preprocessing methods on polyp localization in colonoscopy frames. In: Engineering in Medicine and Biology Society (EMBC), 2013 35th annual international conference of the IEEE. IEEE, Washington, pp 7350–7354
6. Cogal O, Akin A, Seyid K, Popovic V, Schmid A, Leblebici Y (2014) A new omni-directional multi-camera system for high resolution surveillance. In: Proceeding of SPIE defense and security symposium, Baltimore, MD. doi:10.1117/12.2049698
7. Cogal O, Popovic V, Leblebici Y (2014) Spherical panorama construction using multi sensor registration priors and its real-time hardware. In: IEEE international symposium on multimedia (ISM). IEEE, Washington
8. Elahi SF, Wang TD (2011) Future and advances in endoscopy. J Biophotonics 4(7–8):471–481
9. Floreano D et al (2013) Miniature curved artificial compound eyes. Proc Natl Acad Sci 110(23):9267–9272
10. Gluck N, Fishman S, Melhem A, Goldfarb S, Halpern Z, Santo E (2014) Su1221 aer-o-scope™, a self-propelled pneumatic colonoscope, is superior to conventional colonoscopy in polyp detection. Gastroenterology 146(5, Suppl 1):S-406. http://dx.doi.org/10.1016/S0016-5085(14)61467-0, http://www.sciencedirect.com/science/article/pii/S0016508514614670, 2014 {DDW} Abstract
11. Gralnek IM (2015) Emerging technological advancements in colonoscopy: Third Eye® Retroscope® and Third Eye® Panoramictm, Fuse® Full Spectrum Endoscopy® colonoscopy platform, extra-wide-angle-view colonoscope, and NaviAidtm G-EYEtm balloon colonoscope. Dig Endosc 27(2):223–231. doi:10.1111/den.12382. http://dx.doi.org/10.1111/den.12382
12. Gralnek IM, Carr-Locke DL, Segol O, Halpern Z, Siersema PD, Sloyer A, Fenster J, Lewis BS, Santo E, Suissa A, Segev M (2013) Comparison of standard forward-viewing mode versus ultrawide-viewing mode of a novel colonoscopy platform: a prospective, multicenter study in the detection of simulated polyps in an in vitro colon model (with video). Gastrointest Endosc 77(3):472–479. http://dx.doi.org/10.1016/j.gie.2012.12.011, http://www.sciencedirect.com/science/article/pii/S0016510712030647

13. Gu Y, Xie X, Li G, Sun T, Zhang Q, Wang Z, Wang Z (2010) A new system design of the multi-view micro-ball endoscopy system. In: Engineering in Medicine and Biology Society (EMBC), 2010 annual international conference of the IEEE. IEEE, Washington, pp 6409–6412

14. Hasan N, Gross SA, Gralnek IM, Pochapin M, Kiesslich R, Halpern Z (2014) A novel balloon colonoscope detects significantly more simulated polyps than a standard colonoscope in a colon model. Gastrointest Endosc 80(6):1135–1140. http://dx.doi.org/10.1016/j.gie.2014.04. 024, http://www.sciencedirect.com/science/article/pii/S0016510714013923

15. Jeong JK Ki-Hun, Lee LP (2006) Biologically inspired artificial compound eyes. Science 312(5773):557–561

16. Land MF (1997) Visual acuity in insects. Annu Rev Entomol 42(1):147–177

17. Land MF, Nilsson DE (2012) Animal eyes. Oxford University Press, Oxford

18. Lee LP, Szema R (2005) Inspirations from biological optics for advanced photonic systems. Science 310(5751):1148–1150

19. Liu J, Wang B, Hu W, Sun P, Li J, Duan H, Si J (2015) Global and local panoramic views for gastroscopy: an assisted method of gastroscopic lesion surveillance. IEEE Trans Biomed Eng PP(99):1. doi:10.1109/TBME.2015.2424438

20. Lowe DG (1999) Object recognition from local scale-invariant features. In: The proceedings of the seventh IEEE international conference on computer vision, 1999, vol 2. IEEE, Washington, pp 1150–1157

21. Palka J (2006) Diffraction and visual acuity of insects. Science 149(3683):551–553

22. Peng CH, Cheng CH (2014) A panoramic endoscope design and implementation for minimally invasive surgery. In: 2014 IEEE international symposium on circuits and systems (ISCAS), pp 453–456. doi:10.1109/ISCAS.2014.6865168

23. Popovic V, Seyid K, Akin A, Cogal O, Afshari H, Schmid A, Leblebici Y (2014) Image blending in a high frame rate FPGA-based multi-camera system. J Signal Process Syst 76:169–184.doi:10.1007/s11265-013-0858-8

24. Roulet P, Konen P, Villegas M, Thibault S, Garneau PY (2010) 360 endoscopy using panomorph lens technology. In: BiOS. International Society for Optics and Photonics, Bellingham, p 75580T

25. Rubin M, Bose KP, Kim SH (2014) Mo1517 successful deployment and use of third eye panoramic™a novel side viewing video {CAP} fitted on a standard colonoscope. Gastrointest Endosc 79(5, Suppl):AB466. http://dx.doi.org/10.1016/j.gie.2014.02.694, http://www.sciencedirect.com/science/article/pii/S0016510714008645. {DDW} 2014ASGE Program and Abstracts {DDW} 2014ASGE Program and Abstracts

26. Sagawa R, Sakai T, Echigo T, Yagi K, Shiba M, Higuchi K, Arakawa T, Yagi Y (2008) Omnidirectional vision attachment for medical endoscopes. In: The 8th workshop on omnidirectional vision, camera networks and non-classical cameras-OMNIVIS

27. Seo JM, Koo Ki (2015) Biomimetic multiaperture imaging systems: a review. Sens Mater 27(6):475–486

28. Seyid K, Popovic V, Cogal O, Akin A, Afshari H, Schmid A, Leblebici Y (2015) A real-time multiaperture omnidirectional visual sensor based on an interconnected network of smart cameras. IEEE Trans Circuits Syst Video Technol 25(2):314–324. doi:10.1109/TCSVT.2014.2355713

29. Sherk TE (1978) Development of the compound eyes of dragonflies (odonata). III. adult compound eyes. J Exp Zool 203(1):61–79

30. Song YM, Xie Y, Malyarchuk V, Xiao J, Jung I, Choi KJ, Liu Z, Park H, Lu C, Kim RH, Li R, Crozier KB, Huang Y, Rogers JA (2013) Digital cameras with designs inspired by the arthropod eye. Nature 497(7447):95–99. doi:10.1038/nature12083

31. Stanek SR, Tavanapong W, Wong J, Oh JH, De Groen PC (2012) Automatic real-time detection of endoscopic procedures using temporal features. Comput Methods Programs Biomed 108(2):524–535

32. Torre LA, Bray F, Siegel RL, Ferlay J, Lortet-Tieulent J, Jemal A (2015) Global cancer statistics, 2012. CA Cancer J Clin 65(2):87–108. doi:10.3322/caac.21262. http://dx.doi.org/10.3322/caac.21262

33. Uraoka T, Tanaka S, Matsumoto T, Matsuda T, Oka S, Moriyama T, Higashi R, Saito Y (2013) A novel extra-wide-angle-view colonoscope: a simulated pilot study using anatomic colorectal models. Gastrointest Endosc 77(3):480–483. http://dx.doi.org/10.1016/j.gie.2012.08.037, http://www.sciencedirect.com/science/article/pii/S0016510712026582
34. Wang RCC, Deen MJ, Armstrong D, Fang Q (2011) Development of a catadioptric endoscope objective with forward and side views. J Biomed Opt 16(6):066015–066015

Chapter 6
Interconnected Network of Cameras

The Panoptic camera is an omnidirectional multi-aperture visual system which is realized by mounting multiple imaging sensors on a hemispherical frame. Previously, centralized hardware implementation was demonstrated for the Panoptic camera. In this chapter, we present a novel distributed and parallel implementation of the real-time omnidirectional vision reconstruction algorithm of the Panoptic camera. In this approach, new features are added to the camera modules such as image processing, direct memory access, and communication capabilities in order to distribute the omnidirectional image processing into the camera nodes. We present a methodology for the arrangement of camera modules with interconnectivity features into a target interconnection network topology. We give the details of the unique custom-made multiple-FPGA-based hardware platform designed for the implementation of the interconnected network of a 49 camera prototype Panoptic System. Alongside with the system, we present a novel way to represent the omnidirectional data obtained from the Panoptic camera. The presented distributed system can perform faster omnidirectional reconstruction and create higher resolution images in real-time compared to previously presented centralized approaches.

6.1 Introduction

One of the trends in constructing high-end computing systems consists of parallelizing large numbers of processing units. A similar trend is observed in digital photography, where multiple images of a scene are used to enhance the performance of the capture process. This technique is called multi-view imaging and has received increasing attention due to the decreasing cost of digital cameras [9]. Research themes and applications such as increasing image resolution [20], obtaining high

© Springer International Publishing AG 2017
V. Popovic et al., *Design and Implementation of Real-Time Multi-Sensor Vision Systems*, DOI 10.1007/978-3-319-59057-8_6

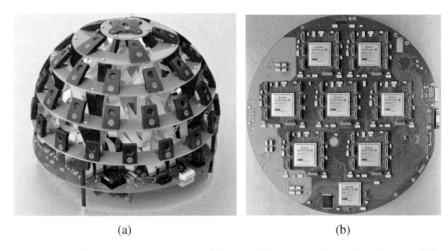

(a)   (b)

Fig. 6.1 (**a**) Built Panoptic prototype with 5 floors and 49 cameras. The sphere diameter of the prototype is $2r_\odot = 30$ cm. (**b**) Top view of the Panoptic Media FPGA-based development platform

dynamic range images [4, 11], object tracking/recognition, environmental surveillance, industrial inspection, 3DTV, and free viewpoint TV [21] are also receiving increasing attention.

Most developed camera array systems are voluminous and not easily portable platforms. Their control and operation depend on multi-computer setups. In addition, image sensors on camera arrays are usually mounted on planar surfaces which prohibits them from covering the full view of their environment. Full view or panoramic imaging finds application in various areas such as autonomous navigation, robotics, telepresence, remote monitoring, and object tracking.

In this chapter, we explain a novel distributed and parallel implementation of the Panoptic camera. Processing workload of the system is distributed among the camera nodes, which aims to overcome the limitations of the centralized approach. To implement the algorithm, a new hardware system is designed and fabricated. Figure 6.1 depicts the new Panoptic Media Platform of 5 floors and 49 cameras utilizing 8 FPGA-based computational board. The new prototype and algorithm presented in this chapter aim to implement the reconstruction algorithm in a parallel and distributed fashion, where image processing applications reside at the camera level.

This chapter is organized as follows. First, we discuss the bottlenecks of the centralized approach and introduce the parallel and distributed approach for omnidirectional vision reconstruction in Sect. 6.2. Then, we talk about the interconnected network of cameras concept and provide guidelines for implementations of such systems in Sect. 6.3. We give details of the designed multiple FPGA-based board in Sect. 6.4 along with the implementation results. Finally, we introduce a novel way of visualizing the omnidirectional data in Sect. 6.5. We will talk about the future prospects of the Panoptic camera in Sect. 6.6. Parts of this chapter were previously published in [5, 18, 19].

6.2 Distributed and Parallel Implementation of Omnidirectional Vision Reconstruction

The system presented in Chap. 4 is implemented using a centralized approach where a single unit is responsible for data acquisition and data processing from multiple image sensors. The real-time implementation of multi-camera applications with a high number of cameras, high image sensor resolutions, and the current image sensor architectures demands a high amount of hardware resources and depending on the target application it might also require high computing performances. This can create bottlenecks in such multi-sensor systems and limits the scalability. The number of cameras that can be connected to a single node is limited by the I/O constraints. For instance, interfacing 49 standard CMOS imagers with a single unit will not be feasible in terms of pin count. Furthermore, for high number of cameras and high camera resolutions, the memory bandwidth requirement increases significantly where a single unit will not be able to sustain the total bandwidth demand. Parallel processing approaches aim to overcome these limitations by distributing signal processing tasks and memory bandwidth usage among several signal processing blocks. This technique creates possibility of constructing higher resolution images beyond the limitations of the centralized approach. Moreover, parallel approaches are faster implementations compared to centralized approaches, which enables the possibility of creating higher resolution images beyond the centralized approach. Due to the constraints posed by technology, the distributed and parallel approach can be a feasible solution for the real-time realization of such systems.

In this chapter, we introduce a novel distributed and parallel implementation of the Panoptic camera. Contrary to the previous systems presented in omnidirectional reconstruction, this novel algorithm and system aim to parallelize the reconstruction algorithm by distributing and parallelizing the tasks to several nodes. The goal is to overcome the physical limitations introduced by the centralized approach. Each individual node is responsible for creating its assigned part of the omnidirectional output frame with the help of the neighboring cameras. In previously published works, a single unit was responsible for creating the whole reconstructed image. This new method allows to increase the number of cameras without additional burden on the reconstruction algorithm, where the centralized approach was strictly dependent on the number of cameras. Furthermore, increasing the number of cameras reduces the amount of workload per node, allowing a higher resolution and higher frame rate solutions.

In this new approach, each node is a single omnidirectional reconstruction system. Thus the features and capabilities of the individual camera nodes must be enhanced which were previously used only for capturing data [1]. Hence, smart camera nodes should possess processing and communication capabilities. The processing capability enables the camera module to perform local processing down to the pixel level, while communication features permit light intensity information exchange among the camera modules. To this aim, a method for creating a regular network topology for camera network is also presented. Converting an irregular topology into a regular topology is conducted in order to generalize the problem, regardless of the source network topology and the camera arrangement in the physical hemisphere dome.

6.2.1 Distributed and Parallel Algorithm

In the distributed and parallel implementation of the Panoptic camera, each camera constructs a portion of the omnidirectional vision with the help of neighboring cameras. For a distributed implementation of the omnidirectional algorithm, each ith camera must possess the knowledge of its covering directions and the information of the other contributing cameras for all of these directions. This information can be extracted by the internal and external calibration processes of the Panoptic System. After extracting the camera parameters, such as camera direction vectors and coordinates on spherical surface and angle of views (AOV) of each camera, each camera can construct its responsible portion of omnidirectional view independently.

For instance, in the nearest neighbor technique, the best viewing camera for each ω is selected. Hence in this technique, each camera constructs a unique set of observation directions. The set of observation directions of each camera has no intersection with the other cameras of the Panoptic System in the nearest neighbor algorithm. Therefore, camera modules can be limited to observe solely their own set of directions and construct their portions of omnidirectional vision, independently from each other.

In the linear interpolation technique, similar to the nearest neighbor technique, each camera can still be assigned to the task of vision reconstruction for its particular partition. For this purpose, each camera would need the information about which other cameras contribute to the particular ω and the intensity values obtained by the contributing cameras. For a constant set of ω directions, these parameters are only required to be calculated once and are stored in a local memory for real-time access. The distributed implementation of the algorithm is summarized in Algorithm 3. The required information can be calculated once by the central unit and updated in the local memory of the camera modules. Alternatively, each camera module can calculate its own required information using its own processing features.

In the initialization process, the set of best observing directions for each camera is extracted. Furthermore, other contributing cameras for each coverage direction and their weights used in the second interpolation step are extracted. After the initialization process, each camera has the knowledge of which ω to construct, which other cameras are contributing to the same ω and, depending on the interpolation type, what are the camera weights contributing to the final level of interpolation. Assuming cameras have processing capabilities, the missing variables to construct the light field are the light intensity values obtained by the other cameras. This creates the necessity of a communication scheme among the camera modules.

Algorithm 3 Distributed reconstruction algorithm for camera nodes

---

1: calculate calibration data
2: calculate weights
3: **for all** best observing directions **do**
4:      $P_m := read\_pixel\_from\_memory$
5:      $p_{contr,2..n} := request\_pixels\_from\_contributing\_cameras$
6:      $C := W_m \cdot P_m + \sum_2^n P_{s,n}$
7:      send C to central unit
8: **end for**
9: **for all** other observing directions **do**
10:      wait for request from principal camera
11:      $P_s := read\_pixel\_from\_memory$
12:      $P_{s,out} := W_s \cdot P_s$
13:      send $P_{s,out}$ to principal camera
14: **end for**

The distributed and parallel implementation of omnidirectional reconstruction algorithm is explained in detail in Algorithm 3. Firstly, the initialization phase is conducted. For each camera, all observing directions (ω) and weights for the chosen interpolation technique are extracted. Then, for each new frame, each camera creates its responsible portion of the final omnidirectional image. For all best observing directions, cameras read from the memory the corresponding pixel light intensity value (P_m) and weight (W_m). In the meantime, the camera module requests contributing light intensity values from the other cameras which observes the same direction. Each camera sends the light intensity value multiplied by the weight. After obtaining all values, the camera sends the sum of all intensity values to the central unit for display.

For directions other than the best observing ones, cameras still possess the weight and light intensity values. When a new light intensity request comes from the best observing camera, the camera reads the light intensity value (P_s) and weight (W_s). Afterwards it reconstructs the light intensity value $P_{s,out}$ for given direction and sends the value to the best observing camera.

6.2.2 Processing Demands

The proposed architecture in [1] performs the omnidirectional vision reconstruction in a pipeline flow for both the nearest neighbor and the linear interpolation techniques. Assuming that the memory used in the system can sustain consecutive access cycles, F_{clk} for the presented real-time omnidirectional vision reconstruction architecture is derived from (6.1) as follows:

$$N_{acs} \times Fps + T_{lat} \leq F_{clk} \qquad (6.1)$$

where N_{acs} is the number of total memory access for reconstruction and F_{ps} corresponds to frame rate. For approximations, the latency term T_{lat} in (6.1) can be neglected. The maximum number of access time is:

$$N_{acs} = N_{cam} \times N_\theta \times N_\phi \qquad (6.2)$$

where $N_\theta \times N_\phi$ is the output resolution, N_{cam} is the number of cameras contributing to image reconstruction in linear interpolation method. The worst case occurs when the cameras contribute in all directions for the linear interpolation technique. Assuming that all data flow occurs in the data processing module with N_{pix} bit presentation, processing F_{clkprc} demands for the proposed centralized system are stated in Table 6.1 for three combination sets of N_{cam}, N_{pix}, camera image frame width (I_w), camera image frame height (I_h), F_{ps}, and different reconstruction resolutions N_θ, N_ϕ. The necessary memory requirements for the whole system, including image frame capturing originated from cameras and omnidirectional vision reconstruction is presented. The aggregate of the latter two demands is translatable into the memory bandwidth requirement of the system using the multiplying factor of N_{pix}. The bandwidth is calculated as follows:

$$(N_{cams}I_wI_h + KN_\theta N_\phi)F_{ps}N_{pix} \qquad (6.3)$$

where K is equal to the number of contributing cameras per ω direction in (6.3). The state-of-the-art current technology for SRAM memories is the Quad Data Rate II SRAM (QDRII-SRAM) with up to 900 million random transactions per second [17]. The maximum data width of a single QDRII SRAM memory is 36 bits [17]. Hence the maximum bandwidth of a state-of-the-art SRAM memory is 26 Gb/s. The bandwidth values that are supported with the state-of-the-art current SRAM technology have a green color background and the ones that are not supported have a red color background in Table 6.1. The table states that in the centralized approach, the increased camera resolution and output resolution will create memory bandwidth problems. Therefore, the distributed and parallel approach will be suitable in order to compete with increasing memory bandwidth and signal processing frequency.

6.2.3   Inter-Camera Data Exchange

In the centralized approach, pixel intensity values were saved in a single unit. Therefore, contributing pixel intensity values are easier to obtain while constructing pixel intensity values of a particular ω direction. However, in the parallel and distributed approach, each camera is responsible for constructing its responsible part of omnidirectional panorama. Assuming that the individual cameras have local processing capability, the only missing variable to construct the light field is the light intensity values obtained by the other cameras. This creates the necessity of a communication scheme among the camera modules. In the next section, we will discuss interconnected network of cameras concept and explain the implemented interconnection scheme among the cameras.

Table 6.1 The clock frequency demand for real-time omnidirectional vision reconstruction (F_{clkprc}) is expressed in MHz

Resolution		N_{cam}	N_{pix}	I_w	I_h	F_{ps}	N_{cam}	N_{pix}	I_w	I_h	F_{ps}	N_{cam}	N_{pix}	I_w	I_h	F_{ps}
		20	8	352	288	25	50	16	640	480	30	100	24	1280	960	60
N_θ	N_ϕ	F_{clkprc}	MBW				F_{clkprc}	MBW				F_{clkprc}	MBW			
64	64	2	0.41				6	7.38				25	176.07			
128	128	8	0.42				25	7.40				98	177.04			
256	256	33	0.45				98	7.50				393	177.32			
512	512	131	0.57				393	7.88				1573	178.46			
1024	1024	524	1.08				1573	9.39				6291	182.99			
2048	2048	2097	3.09				6291	15.43				25,166	201.11			

Implementing the linear interpolation method with $K(n = 4$ cameras per ω). The total memory bandwidth demand (MBW) is expressed in terms of Gb/s. The bandwidth values that are supported with the state of the art current SRAM technology have a green color background and the ones that are not supported have a red color background

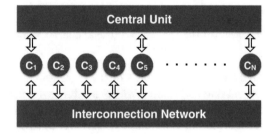

Fig. 6.2 High level model of an interconnected network of cameras. All cameras C_i are connected via interconnection network and some cameras have direct access to central unit

6.3 Interconnected Network of Cameras

An interconnection network is a programmable system capable of transporting data between terminals. The system illustrated in Fig. 6.2 shows N terminals, $C_1 \ldots C_N$ connected to a network. For example, when terminal C_2 wishes to exchange data with terminal C_5, C_2 sends a message containing the data to the network and the network delivers the message to C_5. The terminals C_i resemble the camera nodes with processing and networking features in addition to basic imaging.

We need to keep in mind that having a distributed camera system does not imply the omission of a central unit. For example, a central unit is required for the cameras to send their processed information for the purpose of display. Also a hybrid approach for the application deployment can be considered, where some of the processing is distributed at the camera level and the rest of the processing is conducted in the central unit. For this purpose it is preferred that all cameras also have a direct access to a central unit. Nevertheless, this feature is neither feasible nor optimal in most cases. A central unit may not have enough ports to interface with all cameras of the system. In a case where all the cameras are connected to the central unit with distinct interfaces and the respective bandwidth of these connections are not fully utilized, an inefficient usage of resources is taking place. Hence it is more

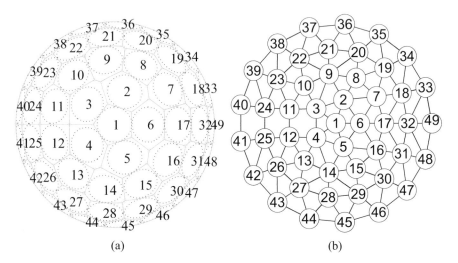

Fig. 6.3 (**a**) Top view of the Voronoi diagram of a five-floor Panoptic System containing 49 camera locations. (**b**) The planar graph extracted from the Voronoi diagram

efficient to provide some of the cameras with direct accessing capability to the central unit and share these connections with the cameras that do not have a direct interface to the central unit. The availability of an interconnection network permits the utilization of this strategy. The latter concept is depicted for the Panoptic System with N cameras in Fig. 6.2.

In multi-camera applications such as the Panoptic camera, information exchange mostly takes part among the neighboring cameras. Thus, during the creation of an interconnection network, neighborhood relations of camera modules should be preserved as much as possible. The neighborhood relation for the cameras in the Panoptic System can be seen in Fig. 6.3b. It is an irregular graph-based topology. However, in most of the systems, this irregular graph-based topology is hard to implement and control on hardware level. A regular graph-based topology can be used to simplify the implementation of the interconnection network. Instead of creating an irregular graph-based network shown in Fig. 6.3b, a regular graph-based, 7×7 mesh topology is chosen in order to realize interconnected network of cameras.

A regular network topology is relatively simple to implement and control. It is scalable and easy to extend, add, or subtract nodes. Flow control mechanisms and packet structures are easier to construct at the hardware level. Furthermore, it generalizes the problem regardless of the source network topology and the camera arrangement in the physical hemisphere dome. However, mapping cameras into the network nodes creates new problems which will be explained in the next section.

6.3.1   Camera Assignment Problem

In order to obtain the neighborhood relation graph of the Panoptic System, the surface of the Panoptic device hemisphere is partitioned into a set of cells centered on the camera locations. Each cell is defined as the set of all points on the hemisphere which are closer to the camera location contained in the cell than to any other camera positions. The boundaries of the cells are determined by the points equi-distant to two nearest sites and the cell corners (or nodes) to at least three nearest sites. This particular partitioning falls into the category of a well-established geometry concept known as the Voronoi diagram (or Voronoi tessellation [3]). The Voronoi diagram of a 5 floors and 49 cameras Panoptic System can be seen in Fig. 6.3a. The geometrical neighborhood relation of 5 floors and 49 cameras extracted from the Voronoi diagram is shown in Fig. 6.3b.

This assignment strategy is known in the context of a facility allocation problem called the Quadratic Assignment Problem (QAP). The QAP models the following real-life problem: In a graph-based topology, for each pair of locations a distance is specified and for each pair of facilities a weight or flow (e.g., the amount of supplies transported between two facilities) is assigned. The problem is to assign all facilities to different locations with the goal of minimizing the sum of the distances multiplied by the corresponding flows. A planar graph representing the neighboring of the cameras is extracted in Fig. 6.3b where the nodes of the extracted graph represent the cameras and its edges resemble the neighboring of the cameras. Hence, in the latter graph two nodes are connected if their respective cameras are geometrical neighbors. The adjacency matrix of this graph can be used as the flow matrix of the QAP.

The QAP is an NP-hard problem; which means there is no known algorithm for solving this problem in polynomial time and even small instances may require long computation time. Among different proposed solutions, sparse version of the GRASP algorithm [14] has given the best result solving the QAP. The assigned camera numbers of Fig. 6.3b are represented on the mesh graph shown in Fig. 6.4a. The assignment allocates the cameras such that all geometrical neighboring cameras are not more than three hops away from each other in the new topology. The number of nodes in the target topology and the cameras of the Panoptic System are the same in the demonstrated example. The same method is applicable if the number of the nodes in the target topology is larger than the number of cameras of the Panoptic System, which assumes to have cameras with no flow exchanges with other cameras. This solution is considered when no regular-based graph topology is selectable to support the exact number of cameras of the Panoptic System.

6.3.2   Central Unit Access

As stated previously, having a distributed camera system does not imply the omission of a central unit. However, we have explained that connecting all units

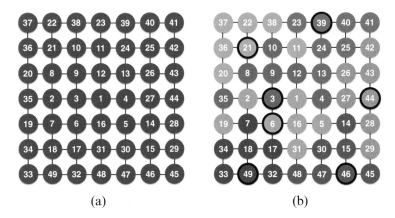

Fig. 6.4 (**a**) The assigned 7×7 mesh topology interconnected network. (**b**) The 7×7 mesh topology with 7 vertex p-centers

to the central unit is also problematic. We need to find candidate cameras that will have direct access to the central unit. In order to decide which cameras will have direct access, the problem to solve is which p candidate cameras to select to have access to the central unit so that the rest of the cameras can access the central unit with minimum number of hops. This feature is desired for reducing the access time between the central unit and any camera of the interconnected network, assuming sufficient channel bandwidth is available. The latter problem can also be mapped into a facility allocation problem known as the vertex p-center problem. The basic p-center problem consists of locating p facilities and assigns clients to them in order to minimize the maximum distance between a client and the facility it is assigned to. This problem is also known to be NP-hard [8]. In order to distribute 49 cameras' load equally, the p value is chosen as 7. As an example, a vertex 7-center problem has been solved for the mesh graph topology depicted in Fig. 6.4b assuming that each camera with access to the central unit can support up to seven clients. The problem is solved using an exact algorithm for the capacitated vertex p-center problem [15]. The solution is depicted in Fig. 6.4b. All the cameras acting as p-center (i.e., with access to central unit) are shown with a bold edge. The cameras belonging to the same p-center are also filled with similar colors. All cameras are at most two hops away from their supporting facility camera. This strategy aims to minimize the network load caused by the transmission of central unit access packets.

6.3.3   Verification

The designed interconnection network is simulated under real or close-to-real conditions. The "BookSim" simulator [7] is used for the purpose of performance analysis of the interconnection network of cameras. The BookSim simulator is a C++ based cycle-accurate interconnection network simulator. The simulator is extended to support custom-defined traffic patterns which are configured by a custom text file. This development was accomplished to support any traffic pattern for target networks under test. A MATLAB-based routine is developed in order to simulate different injection rates with several different test patterns. Optimal parameters for router unit such as number of virtual channels and buffer size are extracted in terms of latency (T_c) versus throughput (λ) with custom created Panoptic traffic pattern. Injection rate is indicating how frequently a new packet is injected into network while latency indicates how many clock cycles it takes for a network packet to traverse to the destination node. All the injection rates are normalized to channel bandwidth and latency is expressed in number of cycles.

The graphs in Fig. 6.5a, b depict the latency vs. injection rate for different number of vertex p-centers selected for direct access to the central unit. It is observed that for the nearest neighbor technique traffic pattern, the demands on the interconnection network tend to reduce as the number of vertex p-centers grows. As the number of vertex p-centers grows, the traffic becomes more balanced and localized.

The 7×7 mesh network is also simulated under linear interpolation traffic pattern. The number of vertex-p centers is chosen as seven. The assignment provided by the QAP approach and shown in Fig. 6.4b is used. The graphs in Fig. 6.5c, d demonstrate the latency versus throughput for routers with flit buffer size equal to 8 and 64, respectively. The results are given for throughput values of $\lambda < 0.4$, as it is expected that the injection rate will not be higher than 0.4.

For the purpose of comparison, a set of average packet latency versus average throughput graphs under linear interpolation traffic pattern for a 7×7 mesh network with random and QAP assigned camera locations. Figure 6.5e, f, demonstrate the latency versus throughput for routers with one virtual channel and flit buffer size equal to 8 and 64, respectively.

Simulations prove that Panoptic traffic pattern can be implemented with expected injection rate and latency. Extracted parameters utilized during the implementation of the router mechanism in an FPGA platform. For the FPGA implementation, an open-source Network-on-Chip Router in RTL provided by [7] is utilized.

6.4 Panoptic Media Platform

A custom-made FPGA platform is designed for the practice of the concept of an interconnected network of cameras. The developed platform is referred to as the Panoptic Media. A Panoptic System comprising 49 cameras is interfaced to this platform. The design and implementation of the parallel and distributed approach of the omnidirectional vision reconstruction algorithm of the Panoptic camera is elaborated for the Panoptic Media platform. The Panoptic Media Board (PMB) is an FPGA-based development board. The PMB includes eight Xilinx XC5VLX110 Virtex5 FPGAs. One FPGA is targeted for the implementation of the central unit and the other seven are slaves and used for emulating an interconnected network

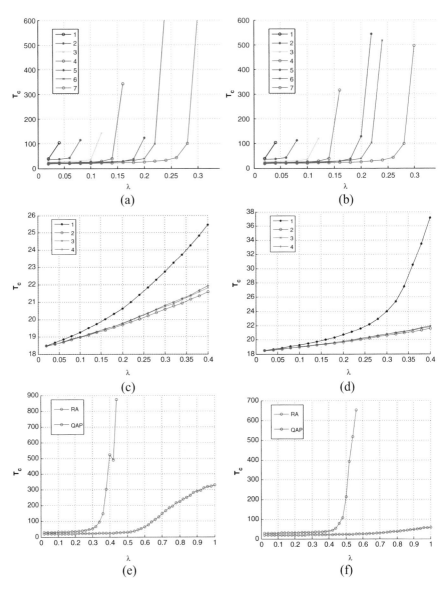

Fig. 6.5 Average packet latency (T_c) vs. average throughput (λ) (i.e., packet injection rate) graphs. The (**a**) and (**b**) graphs demonstrate the latency vs. throughput for routers with flit buffer size equal to 8 and 64. The (**c**) and (**d**) graphs demonstrate the latency vs. throughput for routers for a 7×7 mesh network with QAP assigned camera locations comparing number of virtual channels, flit buffer sizes equal to 8 and 64. The (**e**) and (**f**) graphs demonstrate the comparison in between QAP camera assignment versus random camera assignments, flit buffer sizes equal to 8 and 64. Several different random assignments have been conducted and the average latency values are obtained through BookSim simulations

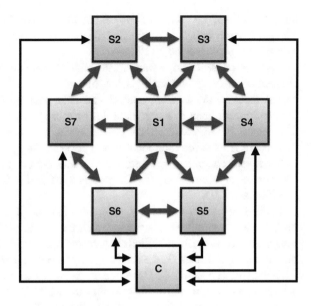

Fig. 6.6 Interconnection scheme between the FPGAs on the Panoptic System

of cameras. The FPGA hosting the central unit is referred to as the central/master FPGA and FPGAs hosting cameras is referred to as the slave FPGAs. The top view of the designed platform is shown in Fig. 6.1b.

Interconnection scheme among the FPGAs can be seen in Fig. 6.6. Thick lines indicate the interconnection between the slave FPGAs and thin lines indicate the interconnect between central unit and other FPGAs. As it can be seen in the figure, not all FPGAs are connected to central unit directly.

6.4.1   Central FPGA

The central FPGA hosts the central unit of the system. It is designed to be in charge of initialization, synchronization among the FPGAs and camera nodes, camera router nodes configuration and control, display and external host communications of the system. For external communications, the central unit has access to a USB-2.0 device and 1 Gb/s Ethernet physical controller device.

At system power up the central unit enters an initialization phase. In this phase, the external physical channel ports of the central FPGA which are connected to that of the slave FPGAs are synchronized. This synchronization is conducted on all FPGAs to achieve a fully synchronous interconnected network. The synchronization is a phase alignment process in which the data bus connections are adjusted at the receiver side for optimum clock sampling. The phase alignment is adjusted using the dynamic time delay adjustment feature of the Virtex-5 FPGA IO buffers. For this purpose, a synchronization pattern is first transmitted on all transmitting bus connections (i.e., outward bus connections) while the receiver bus connections IO buffer time delays are adjusted for optimum clock sampling by their host MicroBlaze processor on all FPGAs.

The central unit can communicate with all camera router nodes of the interconnection network through packet transmission and reception. Two types of packet exist in the system, control and data packets. Control packets are used for configuring camera router modules or monitoring and status check purposes. The central FPGA's MicroBlaze processor can access all the register banks of the SmartCam IPs via the interconnection network using packet-based messages. The data packets contain image information data which are used for display or for transfer to an external host. Each packet type and subtype is identified using a specific packet ID.

Each data packet contains a pixel information of an image frame. The data packets can be simultaneously sent by all the cameras. Therefore the pixels of an image are receivable in a shuffled order by the central unit. Hence all the data packets pertaining to an image frame are first temporarily stored by the RCTRL IP in the ZBT-SRAM. The shuffled order of the receiving data packets implies a random write access nature to a memory. To this aim, the ZBT-SRAM is chosen for temporary storage of the data packets pixel information part. When a full frame is received, the RCTRL IP transfers the received frame to the SDRAM. The SDRAM is used as the video memory for external display interfaces like monitors or projectors.

The central unit has access to a USB-2.0 device through a Xilinx external peripheral controller (EPC) IP. The central unit identifies the USB-2.0 device as an asynchronous FIFO memory. The EPC IP is configured for correct access times with the USB-2.0 device. The USB-2.0 device is used as the primary path for external host communication. The ordered image data in the ZBT SRAM can be transferred to an external host.

6.4.2   Slave FPGAs

The role of a slave FPGA is to emulate a portion of a 7×7 mesh interconnected network of cameras. Each slave FPGA is responsible for seven imagers. Distribution of the camera nodes among the slave FPGAs can be seen in Fig. 6.7. Physical channels between the FPGAs are time-multiplexed, which allows us to create four logical channels between the FPGAs. It can be seen in Fig. 6.7b that there are no more than four channels between FPGAs.

Furthermore, each slave hosts seven camera modules and seven ASRAM memories, application control unit (ACU), and channel synchronization (CHSYNC) modules for inter-FPGA communication and synchronization. The SoC architecture for the slave unit can be seen in Fig. 6.8. Each imager is interfaced to a custom-designed smart camera IP (SmartCam). The SmartCam IP is a camera module with router connectivity, memory, and application processing units. The internal blocks

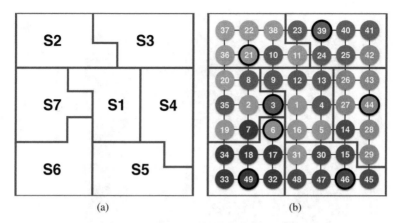

(a)   (b)

Fig. 6.7 Camera assignments of the each FPGA (**a**) Partitioning of the grid based on the assigned FPGA. (**b**) Camera numbers overlayed on the partitions

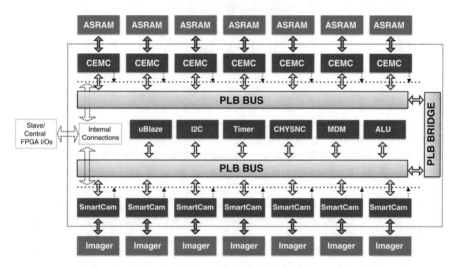

Fig. 6.8 SoC architecture of the slave FPGA

of the custom-designed SmartCam IP are shown in Fig. 6.9. Each SmartCam IP interfaces with a custom external memory controller (CEMC). SmartCam IPs are provided access to an ASRAM via its interfacing CEMC IP.

The SmartCam IP comprises five sub-blocks. The applications intended for the SmartCam IP are implemented into the Image Processing Unit sub-block. The Image Processing sub-block is designed to perform image processing applications. There are three modes of operation, named as video stream, nearest neighbor, and linear interpolation. In the video stream, the SmartCam IP transfers the video stream generated by the camera to the central unit for visual display or external host transfer. This mode is necessary for calibration purposes. In the nearest neighbor and

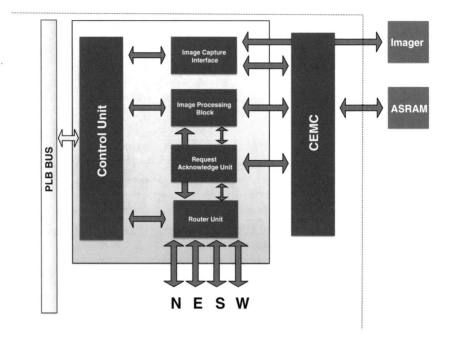

Fig. 6.9 Internal blocks of the SmartCam IP used in the slave FPGAs

linear interpolation modes, it is responsible for creating network demand packets for pixel values obtained by the other contributing cameras and performing the first and second steps of interpolations of the reconstruction algorithm. Each SmartCam IP provides its portion of omnivision of the Panoptic System to the central unit.

The Imager Interface sub-block is responsible for image acquisition and transfers the video stream generated by the imager to the ASRAM memory. The IP sub-block communicates with the central unit and other SmartCam IPs in the Panoptic System through the router sub-block. The router sub-block comprises five-ports (i.e., north, south, east, west, and an input/output port in order to enter or flush out of the network ports). The router sub-blocks' main aim is to create the communication medium among the SmartCams.

The Request Acknowledge sub-block responds to the incoming demand packets from other SmartCam IPs. It creates respond packets that contain the necessary intensity values and coefficients that are used in the second step of interpolation. Control unit sub-block is for the IP's mode configuration, monitoring, and status checks. It can be reached by the central unit via the interconnection network to perform overall control of the system. It organizes the blocks according to current mode of the system.

Forty-nine SmartCams distributed over seven FPGAs are operating in parallel for omnidirectional vision reconstruction. Throughout the interconnection network, pixel intensity values are interchanged among the modules and each camera constructs its assigned portion of the omnidirectional vision. The central unit is responsible for obtaining all reconstructed pixels and displaying them.

6.4.3   Inter-FPGA Communication

Each FPGA has twelve sets of 24-bit bus connections. Each two set of 24-bit bus connections are bundled to form a physical channel port for an FPGA. Each FPGA contains six physical channel ports. The direction of one bus connection is chosen as outward while the other one is selected as inward. However, the physical channel ports of the FPGAs can contain multiple logical channels. For the presented partitioning scheme of a 7×7 mesh interconnected network among the slave FPGAs of the PMB, it is sufficient to have a maximum of four logical channels within a physical channel port. Logical channels are realizable through time multiplexing while operating at higher frequency rates within a single physical channel. Four logical channels are realized by doubling the slave FPGA clock frequency and sending the packets in dual data rate (DDR) mode.

6.4.4   Implementation Results

A Panoptic multi-camera hemisphere of diameter 30 cm is built by stacking circular PCB rings on top of each other as shown in Fig. 6.1a. Each circular PCB ring pertains to one floor of the Panoptic System with five floors. Forty-nine cameras are placed on circular PCBs.

If the integrity check is passed the central FPGAs processor triggers an interrupt event for all slave FPGAs. The slave FPGAs' MicroBlaze processors start programming their interfaced camera modules simultaneously upon receiving the interrupt. The slave FPGAs MicroBlaze processors start initializing the ASRAM memories after programming the camera modules. Once the programming of the ASRAM memories completed, the system starts to display omnidirectional vision in real-time.

The PMB system was found to support the real-time operation of a 7×7 interconnected network of cameras with a 25 fps frame rate, providing an omnidirectional application with an $N_\theta \times N_\phi = 1024 \times 728$ resolution and a linear interpolation method with $K(n = 4; \omega)$ enforcement. An example screenshot from a video can be seen in Fig. 6.10a. For a second operation mode in addition to the XGA output resolution, Panoptic Media is also shown to support a 256×1024 pixel output resolution using nearest neighbor algorithm. During the display of 256×1024 resolution omnidirectional view video, a chosen camera output in VGA resolution can also be displayed below the 360° omnidirectional output. A screenshot from a video of latter can be seen in Fig. 6.10b and an example video can be found in [10].

The slave unit FPGAs are chosen to operate at 108 MHz clock frequency, while the central FPGA operates at 125 MHz. The 108 MHz clock was also chosen to

(a)   (b)

Fig. 6.10 (**a**) Omnidirectional snapshot of resolution 1024×768 for the linear interpolation technique. (**b**) Omnidirectional snapshot of resolution 1024×256 for the nearest neighbor interpolation technique plus VGA (640×480) display of a single selected camera

Table 6.2 FPGA device utilization for central FPGA and an example of the slave FPGAs, for both nearest neighbor technique and linear interpolation techniques

	Central FPGA		Slave NN		Slave LIN		
Resources	Used	Util. (%)	Used	Util. (%)	Used	Util. (%)	Avail.
Slice LUTs	19,495	28	54,360	78	61,416	88	69,120
Slice registers	20,617	29	32,326	47	40,038	57	69,120
BlockRAM/FIFO	93	72	89	69	89	69	128
DSP48Es	3	4	26	40	61	95	64

support the maximum clock frequency (i.e., 27 MHz) of the camera modules. The 125 MHz clock frequency for central FPGA was chosen to support the nominal maximum bandwidth of the 1 Gb/s Ethernet link available on the PMB. Utilization summaries of central FPGA and slave FPGAs are given in Table 6.2.

In order to satisfy real-time requirements of the system, the PMB Platform should construct an omnidirectional output in less than 40 ms, which corresponds to 25 fps. The maximum output resolution equals to 1 million pixels, which is dictated by the reordering memory capacity. The next generation omnidirectional imaging platform that is currently under design will be capable of displaying 4K videos with 30 fps. The average power consumption of the PMB board, when the FPGAs are in the omnidirectional vision reconstruction mode with XGA resolution, was measured as 67.2 W.

6.5   Visualization of the Omnidirectional Data

The Panoptic camera can be used as a perfect example of a telepresence system. Unlike the virtual reality systems, where users are transported to a virtual scene, telepresence allows users to be in another location in real world. Videoconferencing is one example of telepresence. Videoconferencing offers numerous benefits, such as lowering the travel requirements, improving dialog efficiency, and allowing mobility-impaired people to visit distant places. Instead of using narrow angle field of view cameras, we can achieve a better telepresence with the Panoptic camera.

Early omnidirectional imagers were mainly using extreme fish-eye lenses or hyperboloidal mirrors, such as described in [13]. These imaging systems are limited by the resolution capabilities of a single sensor and feature strong distortions. The resolution can be increased by using more modern image sensors, as described in [22], however, the distortions remain. Additionally a big portion of the image is covered by the reflection of the camera lens. In [6] a multi-camera approach is proposed for omnidirectional video generation for telepresence. However, this particular solution cannot be used for real-time video streaming, because the video generation is achieved in post-processing.

In this section, we will present a novel telepresence system, which allows users to naturally observe the remote location. Omnidirectional data will be created with the Panoptic Media Platform and remapping of omnidirectional data through observed direction will be created by using the wide field of view head-mounted display (HMD) Oculus Rift [12].

As explained in Sect. 6.4.4, Panoptic camera has two different operation modes. For the telepresence system, full XGA resolution (1024 × 768) of the [18] will be used. The system can be divided into two parts, one of which is the server application and the second one is the client application.

6.5.1 Server Application

The omnidirectional XGA output generated by the Panoptic camera is transmitted via the DVI output. A capture card connected to the server PC is utilized to transfer omnidirectional data into the server PC. The main task of the server PC is to distribute the whole omnidirectional image via TCP to clients. The application automatically adapts to input resolution changes and can therefore also be used with other camera systems and future versions of the Panoptic camera. The server application is able to stream video to multiple clients at the same time via TCP.

6.5.2 Client Application

The client application receives the TCP stream originated from the server application and generates the views for the head-mounted display. Alternatively, it can also directly receive the images from the DVI capture card, when the camera system is close to the user.

In order to display the hemispherical image on the client side of the telepresence system, a virtual environment is created. This virtual environment is created using the OpenGL API and consists of a user controlled camera and a large overhead hemisphere, onto which the image is mapped. The camera rotates according to the

(a)   (b)

Fig. 6.11 (**a**) The textured OpenGL hemisphere showing a captured image, viewed from the side. (**b**) The client application generating the left and right eye view for the head-mounted display

sensor data received from the head-mounted display. The omnidirectional image is used as a texture for the virtual hemisphere. To retrieve the correct dimensions of the captured objects, the equal density mapping scheme expressed in (3.7) needs to be reversed.

Using the inverse mapping functions (6.4) and (6.5) the original angular directions are restored. In these equations, $\frac{i}{N_\phi}$ and $\frac{j}{N_\theta}$ correspond to the OpenGL texture coordinates s and t, respectively.

$$s(\phi) = \frac{i(\phi)}{N_\phi} = \frac{\phi}{2\pi} \qquad\qquad 0 \le \phi < 2\pi \qquad\qquad (6.4)$$

$$t(\theta) = \frac{j(\theta)}{N_\theta} = 1 - \cos(\theta) \qquad\qquad 0 \le \theta \le \frac{\pi}{2} \qquad\qquad (6.5)$$

Figure 6.11a shows the textured virtual hemisphere from the side. When using the application with the head-mounted display, the user Viewpoint is in the middle of the sphere. Figure 6.11b shows the application in normal use with the HMD.

In order to ensure a high frame rate at all times, the application receives new omnidirectional images in a secondary thread. Thanks to the multi-threaded implementation, the rendering frame rate is independent from the USB or network connection speed, as well as the camera frame rate. This is important for the network streaming functionality, in which the frame rate can vary.

6.6 Conclusion

In this chapter, we presented a novel implementation for omnidirectional vision reconstruction algorithm. Previously, omnidirectional vision reconstructed was conducted in a single unit [2, 16]. In this work, image reconstruction is distributed among the nodes, allowing the addition of more cameras to the system. The method relaxes the I/O constraints and memory bandwidth limitations imposed by the central approach. We have presented an FPGA-based system for implementing the proposed algorithm in real-time. The described system is capable of creating XGA resolution (768 × 1024) of 25 fps real-time omnidirectional video utilizing 49 cameras in parallel. Furthermore, during our tests, we realized that 360° image can be confusing for humans, since mankind is not used to seeing all directions around themselves. Therefore, we have developed a head-mounted display-based visualization system using Oculus Rift. The system renders the omnidirectional view according to the viewer's angle. This allows us to create a telepresence system where each user can separately render his/her own point of view.

We believe Panoptic camera has a promising future ahead and many other research can be conducted using the principal concepts of the distributed approach explained in this chapter.

References

1. Afshari H, Popovic V, Tasci T, Schmid A, Leblebici Y (2012) A spherical multi-camera system with real-time omnidirectional video acquisition capability. IEEE Trans Consum Electron 58(4):1110–1118
2. Afshari H, Akin A, Popovic V, et al (2012) Real time FPGA implementation of linear blending vision reconstruction algorithm using a spherical light field camera. In: IEEE workshop on signal processing systems
3. de Berg M, van Kreveld M, Overmars M, Schwarzkopf O (2000) Computational geometry: algorithms and applications, 2nd edn. Springer, Berlin
4. Debevec PE, Malik J (1997) Recovering high dynamic range radiance maps from photographs. In: Proceedings of the 24th conference on computer graphics and interactive techniques, New York, NY, pp 369–378
5. Gaemperle L, Seyid K, Popovic V, Leblebici Y (2014) An immersive telepresence system using a real-time omnidirectional camera and a virtual reality head-mounted display. In: 2014 IEEE international symposium on multimedia (ISM), pp 175–178. doi:10.1109/ISM.2014.62
6. Ikeda S, Sato T, Yokoya N (2003) Panoramic movie generation using an omnidirectional multi-camera system for telepresence. In: Bigun J, Gustavsson T (eds) Image analysis. Lecture notes in computer science, vol 2749. Springer, Berlin, pp 1074–1081
7. Jiang N, Becker D, Michelogiannakis G, et al (2013) A detailed and flexible cycle-accurate network-on-chip simulator. In: Proceedings of the 2013 IEEE international symposium on performance analysis of systems and software
8. Kariv O, Hakimi SL (1979) An algorithmic approach to network location problems. I: the p-centers. SIAM J Appl Math 37:513–538
9. Kubota A, Smolic A, Magnor M, Tanimoto M, Chen T, Zhang C (2007) Multiview imaging and 3dtv. IEEE Signal Process Mag 24(6):10–21. doi:10.1109/MSP.2007.905873
10. LSM Real-time panoptic video by EPFL-LSM @ONLINE. http://www.youtube.com/user/LSMPanoptic/videos
11. Mann S, Picard RW (1995) On being undigital with digital cameras: extending dynamic range by combining differently exposed pictures. In: Proceedings of IS&T, pp 442–448
12. Oculus VR L Oculus rift. https://www.oculus.com/en-us/
13. Onoe Y, Yamazawa K, Takemura H, Yokoya N (1998) Telepresence by real-time view-dependent image generation from omnidirectional video streams. Comput Vis Image Underst

71(2):154–165. doi:http://dx.doi.org/10.1006/cviu.1998.0705. http://www.sciencedirect.com/science/article/pii/S1077314298907056

14. Pardalos L, Resende M (1994) A greedy randomized adaptive search procedure for the quadratic assignment problem. In: Quadratic assignment and related problems. DIMACS series on discrete mathematics and theoretical computer science, vol 16. American Mathematical Society, Providence, pp 237–261

15. Pinar MC, Ozsoy FA (2006) An exact algorithm for the capacitated vertex p-center problem. Comput Oper Res 33(5):1420–1436. doi:10.1016/j.cor.2004.09.035. http://dx.doi.org/10.1016/j.cor.2004.09.035

16. Popovic V, Afshari H, Schmid A, Leblebici Y (2013) Real-time implementation of Gaussian image blending in a spherical light field camera. In: 2013 IEEE international conference on industrial technology (ICIT), pp 1173–1178. doi:10.1109/ICIT.2013.6505839

17. QDR Consortium (2013) Quad Data Rate SRAM. http://www.qdrconsortium.org/

18. Seyid K, Popovic V, Cogal O, Akin A, Afshari H, Schmid A, Leblebici Y (2015) A real-time multiaperture omnidirectional visual sensor based on an interconnected network of smart cameras. IEEE Trans Circuits Syst Video Technol 25(2):314–324. doi:10.1109/TCSVT.2014.2355713

19. Seyid K, Cogal O, Popovic V, Afshari H, Schmid A, Leblebici Y (2015) Real-time omni-directional imaging system with interconnected network of cameras. In: VLSI-SoC: internet of things foundations, IFIP advances in information and communication technology, vol 464. Springer, Cham, pp 170–197

20. Szeliski R (1994) Image mosaicing for tele-reality applications. In: Proceedings of the second IEEE workshop on applications of computer vision, pp 44–53. doi:10.1109/ACV.1994.341287

21. Tanimoto M, Tehrani M, Fujii T, Yendo T (2011) Free-viewpoint TV. IEEE Signal Process Mag 28(1):67–76. doi:10.1109/MSP.2010.939077

22. Yamazawa K, Takemura H, Yokoya N (2002) Telepresence system with an omnidirectionall HD camera. In: Proceeding of 5th Asian conference on computer vision (ACCV2002), vol 2, pp 533–538

Chapter 7
Towards Real-Time Gigapixel Video

In the previous chapter, we explained the design flow and the full hardware implementation of Panoptic, a real-time omnidirectional multi-camera system. Panoptic is a miniaturized system consisting of fifteen cell phone cameras providing 256×1024 resolution output. In this chapter we will present GigaEye II, a modular high-resolution multi-camera system, capable of achieving gigapixel resolutions.

7.1 Introduction

Panoptic camera presented in Chap. 4 is a scalable system. However, the scalability is reflected in stacking multiple processing boards, and blending the full omnidirectional image at the end. There are three main disadvantages of such approach: (1) each of the stacked boards reconstructs the full panoramic frame, creating a significant data overhead, (2) the resolution of individual cameras is limited due to capacity of the used ZBT SRAMs, and (3) reaching very high resolutions is not possible in real-time.

The problem of data overhead can be solved by using the distributed processing approach instead of the centralized one. Seyid et al. [6] designed an interconnected network of smart cameras, where each camera represents a node in a mesh network, and processes only the pixels in its own FOV. Each camera in this system has a dedicated frame storage SRAM that still limits the maximum resolution of the camera, due to small capacities of static memories. Furthermore, the data traffic patterns and the network latency limit the total system's throughput, effectively lowering the frame rate at high panorama resolution.

In GigaEye II, the aforementioned limitations are resolved as follows:

1. The distributed approach is used by dividing the system into M clusters of N cameras, where each cluster processes only its FOV.

© Springer International Publishing AG 2017   139
V. Popovic et al., *Design and Implementation of Real-Time Multi-Sensor Vision Systems*, DOI 10.1007/978-3-319-59057-8_7

Table 7.1 Main CMV20000 specifications

Parameter	Value
Total pixel array	5124 × 3844
Active pixel array	5120 × 3840
Pixel size	6.4 μm × 6.4 μm
Filter	RGB Bayer color filter
Output format	12-Bit
Data interface	16 LVDS data channels + 1 LVDS control line + 1 LVDS DDR output clock
Frame rate	Up to 30 fps
Sensor control bus	SPI

2. High-capacity double data rate 3 (DDR3) dynamic RAM (DRAM) is used for frame storage within each cluster, allowing scalability on the camera level.
3. Resolution is easily increased by including additional clusters to the system, which are connected either to the central unit, or an intermediate board, via a high-speed link. Thanks to the distributed processing, the processing time is not affected by the addition of the new cluster.

In the rest of the chapter, the full GigaEye II system will be presented in a similar form to Panoptic.

7.2   Camera Module Design

In order to achieve high acquisition resolutions with a reasonable number of image sensors, a CMOSIS CMV20000 color sensor is chosen. The sensor outputs 20 Mpixels at 30 fps frame rate. The summary of the sensor's specifications is given in Table 7.1.

The sensor headboard PCB is designed according to the CMOSIS guidelines, and it can be seen in Fig. 7.1. The sensor is placed on a zero insertion force (ZIF) socket for easy placement and removal. In order to achieve the maximum frame rate, an $f = 480$ MHz clock has to be provided externally. In order to keep the signal integrity for such high frequencies, the high-speed SAMTEC differential cables are used. Furthermore, the power dissipation of the DC/DC converters (marked in Fig. 7.1b) and the image sensors is high, which leads to overheating of the chips, the image sensor, and the PCB. Hence, the used DC/DC converters must be large and with high thermal resistance, and a heat-sink should be placed on the PCB at the back of the sensor.

The analog-to-digital converters (ADC) in the sensor provide a 12-bit digitized value for each pixel that is serialized and sent to the output. The sensor's outputs consist of sixteen low-voltage differential signaling (LVDS) channels, which send sixteen different pixels. The frame is divided into eight vertical strips 640 pixels

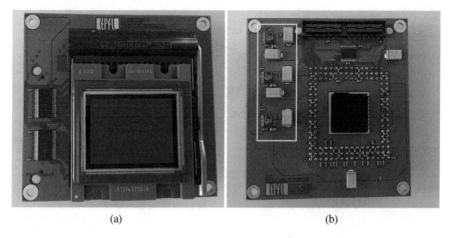

(a)   (b)

Fig. 7.1 (**a**) The front side of the sensor headboard. CMV20000 color sensor is installed on a ZIF socket. Two visible chips are LVDS repeaters to drive the signal through a cable to the processing board; (**b**) The back side of the PCB showing power distribution part marked in a *yellow rectangle*, a SAMTEC connector for multi-gigabit transmission and a heat-sink to cool down the PCB

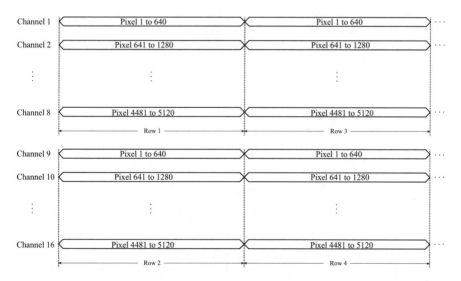

Fig. 7.2 Pixel mapping for sixteen output channels of CMV20000

wide, and two horizontal blocks where one block consists of even rows, and the second one of odd rows. The intersection of a vertical strip and a horizontal block forms one output channel, as shown in Fig. 7.2.

The full camera module is shown in Fig. 7.3a. A lens holder is fabricated using a 3D printer, and it includes an adjustable lens mount cylinder. The cylinder is used to adjust the flange distance, i.e., the distance between the sensor and the mount ring. The flange distance is made adjustable in order to have the flexibility in the

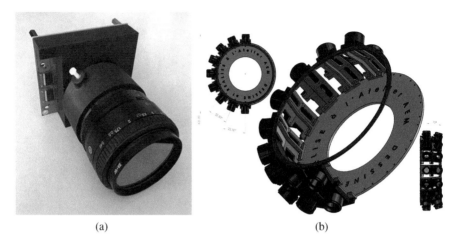

(a)   (b)

Fig. 7.3 (**a**) Assembled camera module with a 50 mm Nikon lens and an external infrared filter, and (**b**) the technical drawing of the GigaEye II structure

choice of lens, image sensor socket, and to compensate for potential imprecision of the 3D printing process. For this camera module, a Nikon F-mount 50 mm lens is chosen. Since, the CMV20000 is the color sensor without any infrared (IR) filtering, an external IR filter is placed on the lens itself.

7.3 System Design

Opposite to the miniaturized Panoptic camera, the camera modules in GigaEye II do not cover the whole hemisphere. A partial 3D model of the GigaEye II structure is shown in Fig. 7.3b. The structure currently allows placement of thirty-two cameras in two rings, with the top ring inclined by 15° upwards. The structure can be upgraded by adding more rings on top of the current one, to construct a full hemisphere if needed. Sixteen cameras are installed and tested in two different arrangements: (1) all sixteen cameras installed on the bottom ring, covering a full 360° view, and (2) sixteen cameras in two rows of eight, with increased vertical FOV.

Multiple FPGA boards are required to process the large amount of incoming data from the cameras. The diagram of the full GigaEye II system is shown in Fig. 7.4. The system is divided into three layers:

1. Cluster boards—Four cameras form a cluster connected to a single FPGA board. The designed processing system inside the FPGA creates a partial panorama and forwards only that part to the layer above.
2. Concentrator board—Four cluster boards are connected to the concentrator board that stores all partial panoramas, and merges them into a single composite frame.

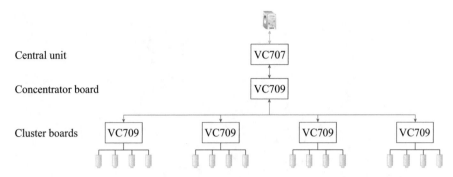

Fig. 7.4 The full system diagram of GigaEye II. The system consists of three main layers: the cluster boards, the concentrator board, and the central unit. The cluster and concentrator boards are XILINX VC709 development kits, and the central unit is VC707. The *red lines* denote the high-speed optical links between the boards, and the *green line* corresponds to the user interface, such as HDMI, USB2, and UART

3. Central unit—The central FPGA board provides an interface towards the user (PC), external displays, and to the concentrator board.

The XILINX Virtex-7 FPGA is chosen to be the main processing core of each layer, since it is the latest generation FPGA providing lots of processing capabilities. The cluster layer and the concentrator layer are implemented on the VC709 development kits. The VC709 board shown in Fig. 7.6a consists of two DDR3 modules, which makes it suitable for the implementation of the image acquisition and image processing algorithms. Similar to Panoptic, during one frame time, one memory module is dedicated to storage of the current frame, whereas the second module is accessed by the panorama construction hardware.

Each cluster board is equipped with an FMC expansion connector. The design of the VC709 board provides 160 user available pins on this connector, which is enough to connect only four CMV20000 cameras. Thus, the design decision on the number of cameras in each cluster is driven by the number of available connections on each FPGA board. An interconnection PCB is designed to connect all four cameras to a single FPGA board, and it is shown in Fig. 7.5. The PCB includes four SAMTEC connectors for camera interface, an FMC connector for the FPGA expansion, and the clock distribution hardware. Since the cameras require a very high-speed $f = 480\,\mathrm{MHz}$ clock that must be transferred through a cable, the strong, low-skew clock drivers are placed on the clock tree for each camera. Furthermore, because of the high data rates, all LVDS lines are length matched, and impedance matched to $Z = 100\,\Omega$.

The connection between the cluster and concentrator boards is realized using the 10 Gb/s optical link that is available on the VC709 board. This link is shown in red color in Fig. 7.4. However, the VC709 board is targeted at intensive processing, and not for user interfacing. Hence, it is not suitable for the implementation of the central unit.

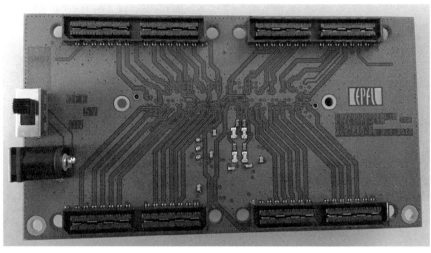

(a)

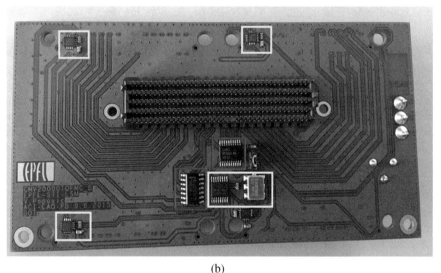

(b)

Fig. 7.5 The FMC interconnection PCB. (a) The top view shows the high-speed SAMTEC connectors used by cameras, and (b) the bottom view shows the FMC connector in the center, and the clock distribution components marked in *yellow rectangles*

The central unit is implemented on a VC707 development kit, shown in Fig. 7.6b. Apart from an optical link interface, the VC707 includes an HDMI port, a USB2 interface, as well as a low-speed serial UART connection. The responsibilities of the central unit include control of the other system boards, indirect control of the cameras via the cluster boards, receiving the full reconstructed panoramic video streams, and providing it on the HDMI output (green line in Fig. 7.4).

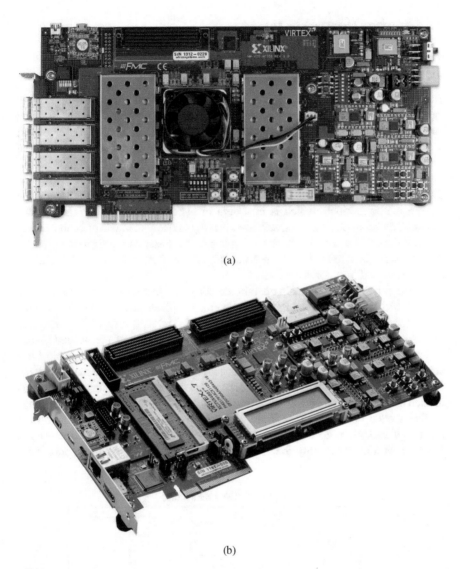

(a)

(b)

Fig. 7.6 Two Virtex-7 development kits used as (**a**) the cluster and the concentrator processing boards (VC709), and (**b**) the central unit (VC707)

Opposite to the centralized architecture implemented on Panoptic, the distributed architecture shown in Fig. 7.4 provides enough processing power to process this amount of pixels in real-time.

The following three sections will give a detailed description of each processing layer, and their internal architecture.

7.4 Cluster Processing Board

7.4.1 Top-Level Architecture

The top-level architecture of the cluster FPGA is depicted in Fig. 7.7. Compared to Panoptic top-level architecture shown in Fig. 4.3, the cluster board is more complex due to several differences between camera modules, such as the raw Bayer output of CMV20000, the high frame resolution, and the different blending method in the Image Processing Unit.

The arrow lines depicted in Fig. 7.7 show the flow of image data inside the FPGA. The serialized image pixels streaming from the cameras enter the FPGA via the camera interface block, which is in charge of synchronizing the FPGA with the cameras, deserializing the data, and multiplexing sixteen channels of each camera. The Raw Image Processing Block performs a de facto standard processing pipeline that includes noise reduction, white balancing, Bayer demosaicing, RGB blending, and contrast and brightness control.

The GigaEye II has two main operation modes. The first one is the full resolution mode, which is similar to the Panoptic camera. All pixels acquired by the cameras are stored in the memory when this mode is used. The reconstruction hardware in the Image Processing Unit generates the memory addresses of the needed pixels, which are then fetched from the memory. While this method is acceptable for a simple panoramic video construction, it is not possible to implement any additional functionality due to random access to DDR3, and the drop in memory performance in such conditions. The performance drop is explained in the following subsection.

The second operation mode is the high-performance mode. This mode allows additional operations to be implemented along the panorama construction, since it estimates and stores only the pixels really needed by the desired application.

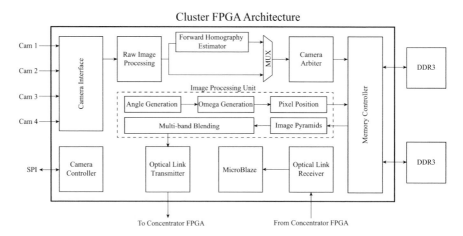

Fig. 7.7 Top-level architecture of the GigaEye II cluster FPGA

Hence, the Image Processing Unit requests the pixels in a stream mode, which is the highest performance mode of any DRAM. The Camera Arbiter implements a time-multiplexing mechanism to store all the incoming frame data from all the camera modules into one of the DRAMs. Similarly to Panoptic, the memory controller interfaces with two external memories on the FPGA board at the same time. The memory controller block provides access for storing/retrieving the incoming/previous frame in/from the DRAMs.

The minor part of the Image Processing Unit rests the same. The angle generation and the omega generation blocks calculate the 3D coordinates of each ω. The angle generation block does not generate angles for the full panoramic view, but only for the selected region observed by the four connected cameras. Unlike Gaussian blending used in Panoptic, GigaEye II implements the MBB algorithm presented in Sect. 3.6. Thus, there is no need to calculate the weights based on the distance of the camera projection from the virtual observer. The pixel position block remains the same.

The MBB algorithm requires decomposition of all four camera frames into LP. The Image Pyramids block is dedicated to that purpose, as well as for generating the GP of weights for each pixel of each camera. Finally, the multi-band blending block merges the four image pyramids into a single one, and sends the data to the concentrator FPGA board via the optical link transmitter.

7.4.2 Camera Interface

The camera interface block deserializes the camera LVDS lines and converts them in a 12-bit parallel pixel data. In order to successfully deserialize the input data, each LVDS channels has to be trained independently. The goal of the training is bit and word alignment of all LVDS channels. The bit alignment is done to ensure that a channel is sampled in the center of its eye diagram. It is achieved by adding delay taps to the input data path of the channel, using the embedded differential input buffers of Virtex-7. The word alignment ensures that the first bit of all channels is sampled at the same clock edge. It is achieved by rotating the received parallel word until the output matches the training sequence.

The goal of the bit alignment is to place the sampling point in the ideal position for each channel. This is done by adding delay taps to the data input line, which shifts it relative to the sampling clock. One bit period consists of two regions: a stable and an unstable region. Sampling in the stable region guarantees that the correct data will always be sampled, regardless of the number of samples that are acquired. In the unstable region, the chance of sampling correct data is not 100%. The unstable region exists due to the non-ideal rise and fall times between two bits, and due to jitter on data and clock lines.

Determining if the sampling point is in a stable or unstable region is done by sampling N 12-bit sequences. If the sampled sequence has N times the same value, the sampling point is considered to be stable. The number of samples N should be high enough for a decent statistical coverage, and it is set to $N = 128$.

Finding the start and end of the bit period is done by continuously determining if a selected sampling point is stable or unstable, adding delay and checking again. At the start of the bit alignment routine, the relative position of clock and data is not known. Therefore, the routine will shift the sampling position until it finds an unstable point. From this point on, the unstable region starts. The training controller will continue shifting the sampling point in the same direction until it finds a stable sampling point. The only remaining point to be found is the end of the stable region. Hence, the controller shifts the sampling point until the next unstable sampling point is found. When this point is found, the controller knows the boundaries of the stable bit period and it can place the sampling point in its center.

The serial data channel contains a continuous stream of bits, and it is impossible for the receiver to know the position of the first bit of a 12-bit word. The goal of the word alignment is finding the position of the first bit in the word. Word alignment is done by continuously sampling the 12-bit training word that the sensor transmits in the training mode. If the sampled 12-bit word does not match the expected training word, the sampling point of the first bit of a word is moved by 1 bit period. This is done until the training word is matched.

Figure 7.8 illustrates the architecture of the camera interface. Apart from the four deserializers, the interface includes channel multiplexers for each camera. CMV20000 outputs sixteen pixels at a time. The pixels are from two consecutive row, with 640 columns offset between them. The goal of 16-to-2 multiplexer shown in Fig. 7.8 is to efficiently reorder the incoming pixels, and create memory addresses for each one of them.

The operation of this multiplexer is driven by parameters of the system. The memory data bus is 512-bit wide. The 12-bit raw pixels are converted into RGB in RGB101010 format, i.e., ten bits are used for each of the color channels. Thus, each pixel can be stored as a 32-bit value, with only two bits of overhead data. These

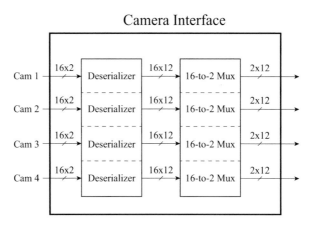

Fig. 7.8 Block diagram of the camera interface block showing deserializers and camera channel time-multiplexers

system parameters allow sixteen pixels to be written to DDR3 at the same time. In theory, it is possible to store the pixels as they come, i.e., to store sixteen arriving pixels. However, this creates a memory addressing problem since the mapping of pixel position to the memory address is non-linear, and requires a resource-demanding hardware.

Hence, a special multiplexer is implemented that buffers eight arriving pixels in each channel. Once eight pixels are buffered, the multiplexer reads sixteen pixels, eight from each row from the same vertical strip. Eight vertical strips are served in a round-robin manner until all the input buffers are read. A memory address jump of 640 pixels is included between each vertical strip. The procedure is repeatedly performed for all pixels in the frame. With this pixel arrangement, the memory addressing is linear, since the neighboring pixels are stored in the adjacent memory addresses. Hence, the hardware that translates the pixel coordinates from Image Processing Unit to the memory address is simple, straightforward, and has low-resource utilization.

7.4.3 Raw Image Processing Pipeline

In general, the Raw Image Processing pipeline is responsible for taking the raw data from the image sensor and generating an image that can be displayed on a screen. Different processing blocks can be included in this pipeline [9] depending on the used camera and its on-chip processing options. The pipeline implemented in GigaEye II for CMOSIS CMV20000 sensor is shown in Fig. 7.9.

The optional downsampler block downsamples the image that comes from the sensor, by removing pixels within a row or a column, in order to achieve the desired output resolution.

The noise reduction block is responsible for reducing the noise produced by the image sensor. The implemented block reduces the noise from three different sources: ADC offset, fixed-pattern noise (FPN), and photo response non-uniformity. Hence, this sub-block is composed of three stages.

In the first stage, a fixed offset defined by a 12-bit input is subtracted from every pixel. This correction essentially sets the dark level of the sensor.

In the following stage, the corresponding 12-bit FPN correction value is subtracted. The FPN is a light-independent noise and corresponds to the standard deviation of an averaged image. It is estimated by taking multiple images in the dark, with short exposure, and averaging them.

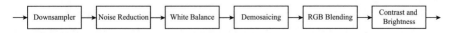

Fig. 7.9 Block diagram of the implemented functions in the Image Processing Pipeline

In the final stage, a gain correction is applied to each pixel in order to correct PRNU. The PRNU is caused by the difference in light sensitivity of the each pixel, i.e., pixels have a different light response curve. It can be obtained by taking several light gray (about 50% of the sensor swing) images, averaging them and subtracting the FPN and the offset.

The final output value of the noise reduction block can be expressed as:

$$I_{corr} = (I_{raw} - I_{offset} - I_{FPN}) \cdot g_{PRNU} \tag{7.1}$$

where I_{corr} is the final corrected pixel intensity, I_{raw} is the raw data from the sensor, I_{offset} is the ADC offset, I_{FPN} is the FPN correction value, and g_{PRNU} is the PRNU gain correction multiplier.

Thanks to correlated double sampling (CDS) done in the pixel, the pixel-to-pixel FPN/PRNU is quite small. More noticeable is FPN/PRNU caused by the column amplifiers on the sensor. Hence, it was chosen to have a per column FPN/PRNU correction, which also reduces memory requirements for storing the correction coefficients. The FPN and gain values are obtained by addressing a BlockRAM in the FPGA, with the column index of the corresponding pixel.

The White Balancing block adjusts the red, green, and blue values so that the white color appears white in the final image, in any lighting condition. This is done by multiplying the red, green, and blue values by different gains factors. The White Balance block is composed of two main parts.

The first part is a circuit that determines the gain values that should be applied to the next frame, based on the values of the current frame. The gains are calculated by computing the means of red, green, and blue, and then dividing them by the smallest mean among the three in order to have gains larger or equal than 1. Calculating the means can be a very intensive process, especially since division is required. The gain estimation is simplified by calculating the mean values among only a subset of pixels. The number of sub-frames, rows, and columns in which the mean is calculated is chosen to be a power of 2, since the dividers can be replaced by a shift right operation. The sub-frames are chosen in such a way that the pixels considered for the mean calculation are well distributed across the image. An illustration of sub-frame distribution is shown in Fig. 7.10. After obtaining the mean value for each color channel, the gains are computed by assuming that the green gain is default 1, and dividing the green mean value by the red and blue means. This is called the "Gray world" method, and in this implementation it provides the color adjustment with respect to the green color.

The second part of the White Balancing block multiplies the gains with the red, green, and blue pixels. This results in the white balanced image, such as the one shown in Fig. 7.11d.

The next block in the processing pipeline is the demosaicing. Since the camera sensor is covered by the Bayer color filter, each pixel receives either red, green, or blue component of light spectrum. Thus, it is necessary to demosaic the image, i.e., to interpolate the missing colors. There are several algorithms for Bayer demosaicing, and the chosen one is as follows. For each pixel, the color components

Fig. 7.10 The distribution of sub-frames considered during white balancing. The *shaded regions* of size 512 × 512 pixels are chosen for an efficient white balancing hardware implementation

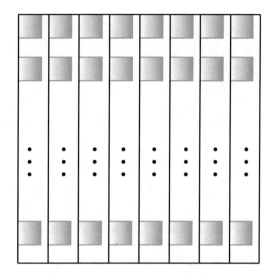

that are filtered out are interpolated from the neighboring pixels having the desired component. The component that is let through the filter remains the same, without considering any of the neighboring pixels. The results of the demosaicing are shown in Fig. 7.11b–d.

The RGB Blending block compensates for the fact that different image sensors produce different RGB values for the same color. Tuning this pipeline stage involves creating a blending matrix to convert the sensor RGB color space to a standard RGB color space. This is done by multiplying each RGB pixel by the matrix obtained by taking images of a calibration ColorChecker chart.

The Contrast and Brightness control is the block that performs a multiplication and an addition/subtraction. First, each color component of a pixel is multiplied by the same contrast coefficient, and then the brightness coefficient is added to each component.

Since the optimal contrast and brightness vary based on the particular lighting conditions, as well as upon user preference, these parameters are implemented so that they are dynamically adjustable by the user.

7.4.4 Forward Homography Estimator

Section 7.4.1 introduced the second operation mode of GigaEye II called high-performance mode. This mode allows implementation of more than one applications, by reducing the memory load and storing only the needed pixels for the desired applications. This is realized using the Forward Homography Estimator (FHE).

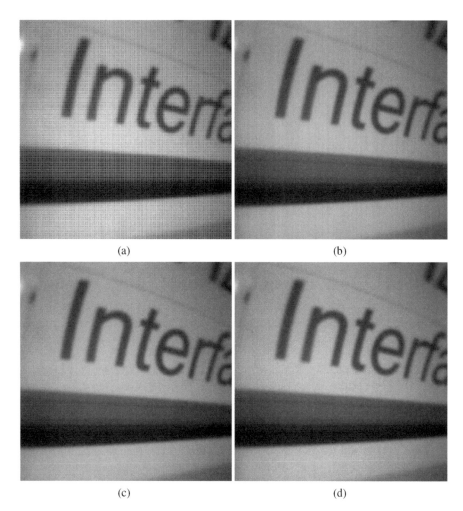

Fig. 7.11 An example of the effect of Raw Image Processing on the final image. (**a**) The raw input image shown in grayscale, (**b**) demosaiced image, (**c**) denoised and demosaiced image, and (**d**) denoised, white balanced, and demosaiced image

Real-time homography is usually perceived as an inverse problem, thanks to a rather simple reconstruction pipeline. It is shown in Sect. 4.5 that the inverse homography is suitable for the application such as panorama construction, since the input images are stored in memory before performing the actual reconstruction. For each desired pixel in the panorama, the most appropriate pixel can be found in the original images. The mapping function is either determined by using runtime calculations [5] or pre-calculated and stored in LUTs [7].

However, timing constraints become very tight when the desired output resolution is high. The image processing systems can hardly meet the real-time constraints of 25–30 fps when reconstructing high-resolution images. Hence, we introduce the forward homography as a possible solution to this problem, which has already been used in stereo image rectification systems [1, 3, 8] where the same real-time constraints apply.

The forward homography solves the issue of the system constraints, such as memory bandwidth, since the correct destination is calculated for each input pixel. Hence, only the necessary pixels for the final reconstruction are stored in memory, thus reducing the required bandwidth. The state-of-the-art forward homography systems pre-calculate the destination coordinates offline, and store them in registers of the processing system.

However, LUT size linearly increases with respect to the input image resolution. CMOSIS CMV20000 sensor outputs 20 Mpixels frames, and LUTs become too large for the FPGA's internal memory. Compressed LUT methods [1] may partially solve this problem, but the peak-signal-to-noise ratio (PSNR) drops significantly in the presence of large differences between input and output image resolution. These differences are also observed in the majority of modern cameras, whose high-resolution images are usually displayed on the 2 Mpixels displays.

Estimating the forward homography in real-time is not a trivial problem. The system should determine the final pixel position in a panoramic image, based only on the pixel coordinates in the original frame. The problem arises due to non-integer values of the mapped pixel coordinates, as illustrated with red dotted lines in Fig. 7.12. When observing homography as an inverse problem, it is easy to scan through the desired pixel grid and choose the closest pixel from the original frame. Forward homography processes a pixel stream, and the system can determine the closest position on the destination pixel grid. However, it cannot determine if the current pixel in the stream is the closest to the destination pixel, since it cannot predict the positions of pixels that have not been processed yet. Thus, pixels that are mapped to the same position are overwritten and the last pixel that appears in the stream will be considered as the correct one. Hence, the PSNR can be significantly

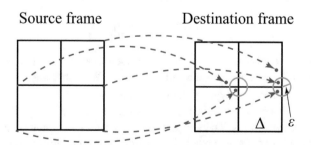

Fig. 7.12 Possible results of the proposed forward homography estimator. Intersections of *black lines* represent source and destination grid pixels, whereas *red lines and dots* are projections and projected pixel locations. The *blue circles* of radius ϵ mark the area in which the projected pixels are considered correct. When more than one pixel is within the *blue circle*, value of the last pixel in the incoming pixel stream is assigned to the pixel in the circle's center

decreased. This problem is even more emphasized in high-performance mode of GigaEye II, where hundreds of pixels from the original 20 Mpixels frame are mapped into a single one in the Full HD panorama.

We developed a new homography estimation algorithm to overcome this issue. By recalling the image formation illustration and equations from Sect. 3.1, Eq. (3.2) expresses the projection of a point in the 3D space onto the image plane. The first step of the algorithm is to back-project the pixels from the image frame. Each pixel $\mathbf{d}$ is back-projected into a line $\mathbf{l}$ in a 3D world that includes the focal point (projection center) $\mathbf{O_c}$. The line is illustrated in Fig. 3.1 and expressed by the inverse of (3.2):

$$\mathbf{l} = M^+ \left[\mathbf{d} \; 1\right]^\top \tag{7.2}$$

where M^+ denotes a Moore–Penrose pseudoinverse of the projection matrix. Thus, back-projection results in a set of lines (light rays), where each one of them contains the focal point of the lens. In order to obtain 3D world coordinates, we should define a back-projection surface. We choose a unit sphere, $|r| = 1$, for the purpose of panorama construction. Back-projection onto the unit sphere is performed by normalizing the line $\mathbf{l}$ by its L^2 norm, and transforming Cartesian (x, y, z) coordinates into spherical (θ, ϕ, r), where θ is the polar angle, ϕ is the azimuth, and r is the radius:

$$\begin{aligned} \mathbf{X}_{\text{sph}} &= \mathbf{l} \, / \, ||\mathbf{l}||_2 \\ \theta &= \arccos\left(\mathbf{X}_{\text{sph}}(z)\right) \\ \phi &= \arctan\left(\mathbf{X}_{\text{sph}}(y)/\mathbf{X}_{\text{sph}}(x)\right) \\ |r| &= 1 \end{aligned} \tag{7.3}$$

If $\mathbf{X}_{\text{sph}}$ is the back-projected pixel, and $\mathbf{X}_s$ is the sampling point on the hemispherical pixel grid, we define a projection error as:

$$e = ||\mathbf{X}_{\text{sph}} - \mathbf{X}_s||_2 \tag{7.4}$$

Afterwards, we find a threshold value ϵ, such that at least one distinct back-projected pixel $\mathbf{X}_{\text{sph}}$ exists for each sampling point $\mathbf{X}_s$ with the error $e \le \epsilon \le \frac{\Delta}{\sqrt{2}}$, where Δ is the distance between two pixels on the hemisphere.

Five outcomes are possible in a 2D homography depending on the source and destination pixel positions. The simplest one is a 1-to-1 mapping when each pixel from the source frame maps to one in the destination frame. Furthermore, 1-to-0 and 0-to-1 are also possible, when the source pixel does not have a corresponding pixel in the destination frame, and vice versa. These three mappings are trivial cases and they will not be considered in the analysis.

Complex 1-to-N and N-to-1 mappings occur when destination and source frames are oversampled, respectively. We resolve the 1-to-N mapping in the estimation algorithm by choosing the optimal ϵ. The optimal ϵ value ensures a distinct source pixel for each ϵ-neighborhood in the destination frame. Hence, a 1-to-N mapping is replaced by N 1-to-1 mappings.

Oppositely, ϵ value should be kept as low as possible in order to efficiently resolve the N-to-1 mappings. The problem that arises in forward homography is that the Nth pixel in the stream is considered as the correct, unless a full mapping is stored in the internal LUT or the external memory. Thus, the optimal ϵ is the lowest value that ensures 1-N resolving. Minimizing the ϵ value in the proposed estimation also reduces the number of pixel candidates to $M < N$. The benefit of this reduction is two-folded: (1) increased chance of choosing the optimal pixel, and (2) smaller error and higher PSNR when non-optimal pixel is chosen.

The homography between the image frame and unwrapped hemispherical surface is shown in Fig. 7.12. Intersections of black lines represent pixel positions on the respective grids. Red dashed lines illustrate homography between two frames, and red dots are projected pixel positions in the destination frame. Distances between red points and the closest intersection of black lines is the corresponding error e. The blue circles mark the ϵ-neighborhood in which projected pixels are considered as potential candidates for the final pixel value.

Figure 7.12 illustrates two different cases of N-to-1 homography. Two source pixels on the left are mapped to the vicinity of a single destination pixel. The ϵ-neighborhood around the central destination pixel is set such that only one of the mapped pixels is inside the circle. Hence, the central destination pixel is given the value of the pixel inside the circle, which is indeed the closest projected pixel. In another situation, three pixels on the right side of the source frame are projected around one destination pixel. Two projections are inside the ϵ-neighborhood and one of them will be chosen as the destination pixel, i.e. the last one read out from the sensor.

Internal architecture of the FHE is shown in Fig. 7.13. Subtraction of the camera center point position (x_0, y_0) translates the image frame origin to the frame center. Different row vectors of the matrix M^+ are provided to the dot product blocks, which evaluate the matrix multiplication in (7.2). A single dot product block in Fig. 7.13 is implemented as a pipelined multiply-accumulate unit in order to increase the performance of the system.

Equation (7.3) expresses the hemispherical back-projection and coordinate system change from Cartesian to spherical. The L^2 norm sub-block in Fig. 7.13 consists of two consecutive square root calculations. The square root module implements a CORDIC algorithm in the vectoring mode, which calculates the L^2 norm of its two inputs, i.e., $\sqrt{a^2 + b^2}$. The dividers for coordinate normalization are implemented using the iterative fast Anderson algorithm [2]. Transformation of the coordinate system requires evaluation of the inverse trigonometrical functions arctan and arccos. The spherical angles (θ, ϕ) are calculated by applying the CORDIC algorithm to the Cartesian coordinates, as illustrated in Fig. 7.13.

The angle generation block provides information on the desired pixel grid. The error e from (7.4) is evaluated by the identical square root module used for the previous L^2 norm calculations. The error is compared to the pre-calculated ϵ,

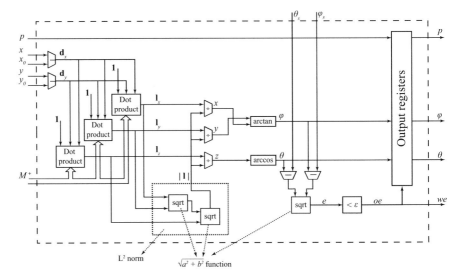

Fig. 7.13 Internal architecture of Forward Homography Estimator. The presented hardware evaluates expressions (7.2)–(7.4) using pipelined architecture. Pipeline registers are not shown for better visibility. The sub-blocks for square root and trigonometric functions evaluation utilize the CORDIC algorithm, whereas the fast Anderson algorithm [2] is used for implementation of the dividers

which is calculated by MicroBlaze using the camera calibration data. Output of the comparator serves as the output enable signal for a set of output registers, and as a write enable signal for the Camera Arbiter in Fig. 7.7.

The FHE is a fully pipelined block. Each computation is followed by a register to shorten the critical path and increase the maximum frequency. Furthermore, each sub-block, e.g., dot product, square root, and trigonometric functions, is also pipelined providing a very fast operation. The pipeline registers are not shown in Fig. 7.13 for clarity reasons.

7.4.5   Image Pyramids

Whether FHE is used or not, the Image Processing Unit operates identically. The pixel position module generates (x, y) coordinates of the pixel in ω direction and fetches it from memory. GigaEye II system implements distributed architecture of MBB, and the core processing part is generation of the multi-resolution pyramids, as shown in Fig. 7.14.

Apart from requesting pixels from the external memory, the pixel position informs the Image Pyramids block about the camera index, weight, and validity of the requested pixel. These three signals form a pixel descriptor, which is stored

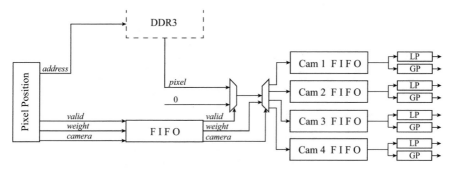

Fig. 7.14 The block diagram of the Image Pyramids processing block. The block interfaces with both the pixel position block and the external memory. The block demultiplexes the pixel data with the camera index as a select signal, and sends it to the appropriate LP and GP decomposition circuit

in the FIFO while the pixel is being read from the external DRAM. The weight can either be 1 if the camera is the best observing one for the selected pixel, or 0 otherwise. These weights correspond to the lowest level of the image LP, i.e., the high-frequency content as explained in Sect. 3.6. The *valid* signal indicates if the pixel is in the FOV of the camera. If the pixel is not in the FOV of the camera, the value 0 (black pixel) is sent to the corresponding FIFO instead of the pixel value read from DRAM.

The pixel value and the weight are demultiplexed using the camera index as the select signal, and stored in one of the four FIFOs corresponding to the cameras in the system. The purpose of these FIFOs is to synchronize LP and GP decomposition. The FIFOs do not output pixels to the LP and the GP blocks until all four FIFOs have at least one stored pixel. This synchronization guarantees that image pyramids are created at the same time and the following blending block can safely implement weight multiplication.

The LP decomposition follows the algorithm illustrated in Fig. 3.12 [4]. The same principle is applied to the GP decomposition, but without the downsampling and interpolation with **G(z)**. Both image quality and timing performance are dependent on the filters **H(z)** and **G(z)**. The 2D FIR filters are often used in FPGA and ASIC designs due to their inherent stability and simplicity of design in digital systems. The implementation of 2D FIR filters can be either separable or non-separable. The non-separable (direct) implementation consists of a 2D convolution of the filter matrix with the image. For $N \times M$ image resolution and $K \times K$ filter matrix size, the computational complexity of such filtering is $\mathcal{O}(MNK^2)$. The hardware design requires K^2 multipliers and a complex input buffer structure for larger filter sizes.

Oppositely, separable filters require less multipliers and adders compared to the direct implementation. However, the traditional separable implementation based on software algorithms is very resource-demanding and quite inefficient. Such computation is mathematically expressed as:

$$x' = (x * h_r)^T * h_c \qquad (7.5)$$

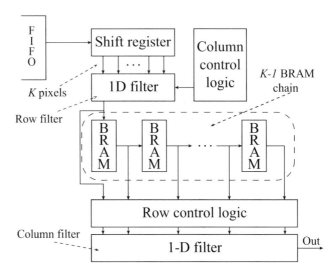

Fig. 7.15 Internal architecture of the 2D separable filter. Both analysis and synthesis filters are implemented using the same architecture, with a slight difference in the control logic blocks

where x and x' are the original and the filtered image, respectively, and h_r and h_c are row and column 1D filters. Operation denoted with $*$ represents a 1D convolution. Without loss of generality, in the rest of the book we will consider symmetric 2D filters, i.e., $h = h_r = h_c$.

The main issue of this implementation is the transposition block. Even though the complexity of $\mathcal{O}(MNK)$ is lower compared to the direct filtering, it requires more memory, as the whole intermediate image result is buffered. The buffering is obligatory due to reordering (transposing) of the pixels before the second 1D filter is applied, which increases the system latency by $N \times M$.

The internal architecture of the analysis and synthesis filters is shown in Fig. 7.15. The goals of this design are real-time performance and reduction of required hardware in the transposition block. Real-time performance is achieved by reducing the critical path delay using pipeline architecture, i.e., result of each arithmetic operation in the algorithm is followed by a register. Hence, the critical path is reduced to the length of the longest path in a single arithmetic block. Furthermore, the proposed design includes data sharing, which reduces number of memory read requests, increases performance, and reduces hardware complexity.

7.4.5.1 Analysis Filter

The analysis filter shown in Fig. 7.15 operates as follows. The pixels that are ordered row-wise are read from one of the camera FIFOs. K pixels are buffered in a shift register, where K is the length of the used 1D filter. The pixels in the register are shifted with the arrival of each new pixel. All K pixels are available at the output and they are used by 1D row filter.

The row filter provides horizontally filtered pixels at its output. In standard separable filter implementations, these filtered pixels are stored in memory, transposed, and filtered again. However, using the proposed architecture, we avoid storing and transposing the full frame. The intermediate memory is replaced by a chain of $K-1$ line buffers, which are implemented as BlockRAMs in the FPGA.

Furthermore, not all filtered pixels are needed in the subsequent stages, because of the downsampling in the LP algorithm. Hence, we introduce two new blocks, named *Column control logic* and *Row control logic* in Fig. 7.15. Since the filtered image will be downsampled, we distribute the downsampling operation into row and column procedures, and embed it in the hardware filter. When one pixel is filtered by the row filter, the *Column control logic* disables the filtering of the next pixel, i.e., pixel positioned in the next column. After skipping one pixel, the control logic again enables the filter. This principle is repeated for all pixels in the image, and it corresponds to the horizontal downsampling by two.

The pixels belonging to the same row are buffered in the same BlockRAM, and only $K-1$ half-rows are stored in this chain, thanks to the control logic. Whenever a new filtered pixel arrives, it is stored in the first BlockRAM at the location addressed by the pixel's column in the frame. Since the utilized BlockRAMs behave as a dual port memory, the second port is used for reading the pixel from the same memory location, i.e., the pixel in the same column from the previous row. The read pixel is then stored in the following BlockRAM in the chain. Hence, this BlockRAM chain can also be regarded as a set of stacked shift registers.

The outputs of $K-1$ BlockRAMs and the output of the row filter form a set of K vertically neighboring pixels. Hence, the transposition is no longer required, as the pixels are available in the appropriate order. Similar to *Column control logic*, *Row control logic* block disables filtering of every second row in the column filter. It is important to note that even when column filtering is disabled, shifting of pixels between BlockRAMs is enabled. This is obligatory due to the fact that one source pixel contributes to $(K+1)/2$ filtered pixels in a single column.

The pixels allowed through the *Row control logic* are filtered using the second 1D filter (column filter in Fig. 7.15) and streamed out to the rest of the processing system. The outputs are sorted in the same order as the original input, i.e., in the row-wise order.

7.4.5.2 Synthesis Filter

Opposite to the analysis filter $\mathbf{H(z)}$ that downsamples the image, the synthesis filter $\mathbf{G(z)}$ upsamples it. A property of the upsampling operation is that output data rate of the filter is higher than the input data rate. We implement a time-multiplex system to resolve this issue, under the safe assumption that the blending operation does not increase the data rate.

The synthesis filter is implemented using the same top-level architecture as the analysis filter. The main difference is in the control logic blocks of the synthesis filter, since it multiplexes the input pixels with the upsampled zero-valued pixels.

When the filter receives a pixel from ith column, the *Column control logic* allows the row filter to output pixel from $i-(K-1)/2$ column. In the following clock cycle, the logic will enable the filter to output the $i-(K+1)/2$ column. The insertion is allowed because of two reasons: (1) the corresponding input pixel for the second output pixel is zero, and (2) the assumption that input data rate is not faster than the output rate of the LP decomposition. Levels l_i, for $i = \{2, \ldots, L\}$ cannot provide pixels in each clock cycle, hence upsampling of the pyramid levels can be embedded in the filtering operation. Level l_1 is the only level that can theoretically provide pixels every cycle, but its pixels are not being filtered by $\mathbf{G(z)}$ during the reconstruction (see Fig. 3.12).

The line buffers store the upsampled rows. The *Row control logic* operates on the same principle as *Column control logic* with the exception that it inserts the row pixels. When the column filter provides a pixel from ith row, the *Row control logic* enables the column filter to output the pixel from $i-(K-1)/2$ row. In the following clock cycle, a pixel from $i-(K+1)/2$ row will be calculated.

The outputs of the synthesis filter are the levels of LP, and they are the inputs of the Image Blending block.

7.4.6 Image Blending

The hardware implementation of the Image Blending block is depicted in Fig. 7.16. Similarly to Panoptic, the Image Blending module in GigaEye II conducts the final step of the partial panorama construction. The module receives frames from four

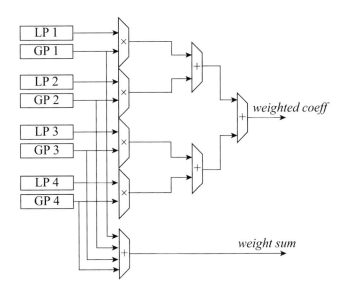

Fig. 7.16 Block diagram of the Image Blending module in the cluster FPGA of GigaEye II

cameras decomposed into LP, and their corresponding weight GP. The coefficients from image LP are multiplied by the weights from the GP. Furthermore, weights from all four GPs are summed to obtain the normalizing factor, as in (3.12). However, the weighted LP coefficients are not normalized at this moment, since the final blending is performed in the concentrator FPGA. As the blending result in the cluster FPGA, the weighted LP coefficients, and sum of all corresponding weights are sent to the concentrator FPGA via the optical link transmitter.

7.5 Concentrator Processing Board

The concentrator processing board (Fig. 7.17) collects data from the four cluster boards. The Memory Arbiter block behaves in the identical manner as the arbiter in the cluster FPGA, i.e., four input channels are served in the round-robin pattern. The DDR3 memory mapping is also identical to the one in the cluster FPGA, with each camera stream replaced by the stream coming from the cluster FPGA. The main difference between the cluster and the concentrator FPGA design is in the Image Processing Unit that is simplified. There is no need for the angle or ω generation, since there is no direct access to the camera frames. The pixels are already arranged in a sequential order. Thus, a counter is instantiated in the pixel position block, whose value is linearly mapped to a memory address.

Recall that the cluster boards calculate and send both the weighted LP coefficient and the sum of corresponding weights. These values are stored in DRAM of the concentrator board, and read by the Image Processing Unit. The MBB block sums all weighted LP coefficients, sums all weights from GP, and normalizes the weighted sum, as shown in Fig. 7.18. The resulting value is the blended LP coefficient of the final panorama.

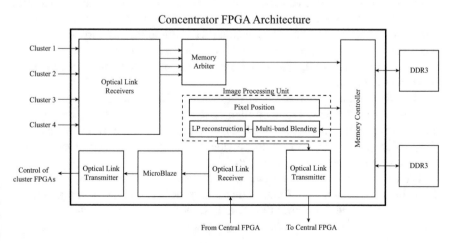

Fig. 7.17 Top-level architecture of the GigaEye II concentrator FPGA

Fig. 7.18 Block diagram of
the Image Blending module
in the concentrator FPGA of
GigaEye II

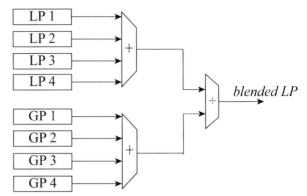

The final panoramic frame is reconstructed from the blended LP coefficients using the chain of interpolation filters, as shown in Fig. 3.12. The reconstructed panorama is transmitted to the central board via optical link.

Apart from the data path, the concentrator boards also include the system control path. The commands from the central FPGA are decoded, and interrupt is raised in the MicroBlaze. The interrupt routine is serviced, and the command is forwarded to the cluster FPGA if needed.

7.6 Central Processing Board

The top-level architecture of the central FPGA is shown in Fig. 7.19. The central FPGA is connected to the concentrator board via 10 Gb/s optical link. The Optical Link Receiver receives the already constructed panoramic frame, generates the valid memory address, and stores it into a single DDR3 module on the board. The memory dedicated to video streaming is divided into two pages, where one frame occupies one page. While the current frame is being written into the memory, the previous one is read by the display controller. Thus, it is not possible to change the contents of the frame being displayed, which leads to possible image "tearing" in the presence of fast moving objects.

The display controller consists of five instantiated HDMI controllers, allowing five independent outputs. The controllers have direct memory access, and read the already stored full frames. Since the VC707 board has only one HDMI output, two FMC extension boards are used to provide additional display capabilities. These extension boards are shown in Fig. 7.20.

The central FPGA is also used as an interface towards the user, i.e., a client PC. It has both USB2 and UART connectors. Since there is no high data rate transfer between the PC and the central FPGA, only UART link is used to send commands from the PC. The user can directly set any software accessible register in the central

Central FPGA Architecture

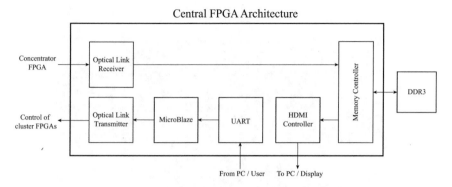

Fig. 7.19 Top-level architecture of the GigaEye II central FPGA

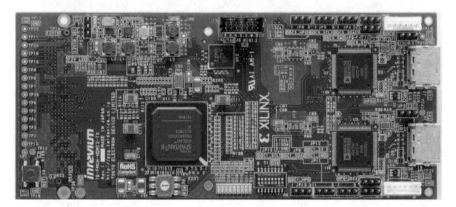

Fig. 7.20 FMC extension board providing two additional HDMI outputs

FPGA through MicroBlaze interrupt routine. Furthermore, it can indirectly set any concentrator and cluster FPGA register, as well as any camera register, since the commands are propagated via the Optical Link Transmitter.

7.7 Experimental Results of the GigaEye II System

The camera mount ring for GigaEye II with diameter of $2r = 50$ cm is built using a 3D printer. The camera ring can currently accommodate thirty-two CMOSIS CMV20000 cameras, arranged on two floors. The hemisphere populated with cameras is positioned on top of an aluminum rack that holds six processing FPGA boards and the power supply. The whole structure is shown in Fig. 7.21. The camera modules are connected to the FMC interface PCB using the high-speed SAMTEC Q Rate differential cables, and they are operated at 30 fps.

Fig. 7.21 The mounted cameras on the GigaEye II camera system

The architecture presented in this chapter was developed in VHDL for the target FPGAs. Similarly to Panoptic, the developed firmware conducts all mathematical processing using 16-bit fixed-point precision. The firmware was targeted and successfully tested for operation at 233 MHz f_{clk} frequency on the cluster and concentrator FPGAs, and 200 MHz on the central FPGA. These frequencies are chosen since they are the frequencies of the on-board oscillators. The resource utilization of each board separately is shown in Table 7.2.

Due to the limits of the modern-day displays, the final real-time output resolution is 1080p30, i.e., 1920×1080 pixel at 30 fps. However, thanks to 20 Mpixels cameras and the adaptable reconstruction algorithm, it is again possible to reconstruct a narrower FOV while keeping the same output resolution, providing more details in that area.

7.8 Conclusion

In this chapter, a real-time high-resolution multi-camera system GigaEye II is presented. GigaEye II is implemented using the distributed processing approach. The full design was detailed, including the camera choice and specifications, the PCB design, and the multi-board real-time hardware implementation of the omni-directional view construction. The hardware resource utilization of all processing FPGAs in the system is also provided.

Table 7.2 GigaEye II Virtex-7 FPGA resource utilization summary

Resource	Cluster	Concentrator	Available	Central	Available
Flip-flop	118,246	24,902	866,400	21,403	607,200
LUT	79,575	25,614	433,200	23,423	303,600
BlockRAM	386	112	1470	81	1030
DSP	131	8	3600	6	2800
MMCM	7	4	20	4	14
PLL	2	4	20	1	14
BUFG	18	12	32	12	32

We presented one possible solution to the problem of designing a high-resolution real-time multi-camera system. The distributed approach for implementing real-time applications in multi-camera systems is very efficient in terms of processing power and speed, but not easy to design and synchronize. Thanks to the workload distribution and parallel implementations, this system achieves the high resolution and the high frame rate operation. The presented embedded architecture with the true distributed workload, i.e., each board reconstructing only a partial FOV is novel in the field, and provides numerous application possibilities.

Finally, the system-level architecture of GigaEye II with the cluster and the concentrator boards allows straightforward scalability of the system. Addition of new four-camera clusters does not require any change in system's architecture or the used algorithm, thanks to fully distributed reconstruction. Hence, GigaEye II can easily reach the gigapixel resolutions with enough number of installed cameras.

References

1. Akin A, Baz I, Gaemperle L, Schmid A, Leblebici Y (2013) Compressed look-up-table based real-time rectification hardware. In: IFIP/IEEE 21st international conference on very large scale integration (VLSI-SoC), pp 272–277. doi:10.1109/VLSI-SoC.2013.6673288
2. Meyer-Baese U (2007) Digital signal processing with field programmable gate arrays, 3rd edn. Springer, Berlin
3. Park DH, Ko HS, Kim JG, Cho JD (2011) Real Time rectification using differentially encoded lookup table. In: Proceedings of the 5th international conference on ubiquitous information management and communication. ACM, New York, pp 47:1–47:4
4. Popovic V, Leblebici Y (2015) A low-power 490 MPixels/s hardware accelerator for pyramidal decomposition of images. In: Proceedings of IEEE international conference on image processing (ICIP), Quebec City, QC
5. Popovic V, Afshari H, Schmid A, Leblebici Y (2013) Real-time implementation of Gaussian image blending in a spherical light field camera. In: Proceedings of IEEE international conference on industrial technology, pp 1173–1178. doi:10.1109/ICIT.2013.6505839
6. Seyid K, Popovic V, Cogal O, Akin A, Afshari H, Schmid A, Leblebici Y (2015) A real-time multiaperture omnidirectional visual sensor based on an interconnected network of smart cameras. IEEE Trans Circuits Syst Video Technol 25(2):314–324. doi:10.1109/TCSVT.2014.2355713

7. Szeliski R (2011) Computer vision: algorithms and applications. Springer, New York. doi:10.1007/978-1-84882-935-0
8. Vancea C, Nedevschi S (2007) LUT-based image rectification module implemented in FPGA. In: IEEE international conference on intelligent computer communication and processing, pp 147–154. doi:10.1109/ICCP.2007.4352154
9. Zhou J (2007) Getting the most out of your image-processing pipeline. Tech. rep., Texas Instruments

Chapter 8
Binocular and Trinocular Disparity Estimation

Depth is a strong component of human vision. The stereoscopic imaging and utilization of glasses can provide depth perception to the user without the requirement of depth measurement. However, its measurement is required for many recent advanced virtual reality applications and 3D-based smart vision systems. Depth estimation is an algorithmic step in a variety of applications such as autonomous navigation of robot and driving systems [20], 3D geographic information systems [23], object detection and tracking [7], medical imaging [8], computer games and advanced graphic applications [22], 3D holography [11], 3D television [15], multi-view coding for stereoscopic video compression [14], and disparity-based rendering [17]. These applications require high accuracy and speed performances for depth estimation. This chapter first introduces disparity estimation. Afterwards, binocular and trinocular disparity estimation algorithms and their hardware implementations are explained.

8.1 Binocular Adaptive Window Size Disparity Estimation Algorithm and Its Hardware Implementation

A hardware-oriented adaptive window size disparity estimation (AWDE) algorithm and its real-time reconfigurable hardware implementation are presented in this section. The implemented hardware processes high-resolution (HR) stereo video with high-quality disparity estimation results [1]. In addition, the disparity estimation quality of the AWDE algorithm is improved using the iterative disparity refinement process. The proposed enhanced AWDE algorithm that utilizes Iterative Refinement (AWDE-IR) is implemented in hardware and its implementation details are presented [2].

© Springer International Publishing AG 2017
V. Popovic et al., *Design and Implementation of Real-Time Multi-Sensor Vision Systems*, DOI 10.1007/978-3-319-59057-8_8

8.1.1 Binocular Hardware-Oriented Adaptive Window Size Disparity Estimation Algorithm

The main focus of the AWDE algorithm is its compatibility with real-time hardware implementation while providing high-quality DE results for HR. The algorithm is designed to be efficiently parallelized, to require minimal on-chip memory size and external memory bandwidth.

The term "block" is used in this chapter to define the 49 pixels in the left image that are processed in parallel. The term "window" is used to define the 49 sampled neighboring pixels of any pixel in the right or left images with variable sizes of 7×7, 13×13, or 25×25. The pixels in the window are used to calculate the Census and BW-SAD cost metrics during the search process.

The algorithm consists of three main parts: window size determination, disparity voting, and disparity refinement. The parameters that are used in the AWDE algorithm are given in Sect. 8.1.4.

8.1.1.1 Window Size Determination

The window size of the 49 pixels in each block is adaptively determined according to the mean absolute deviation (MAD) of the pixel in the center of the block with its neighbors. The formula of the MAD is presented in (8.1), where $\mathbf{c}$ is the center pixel location of the block and $\mathbf{q}$ is the pixel location in the neighborhood, N_c, of $\mathbf{c}$. The center of the block is the pixel located at block(4, 4) in Fig. 8.1. A high MAD value is a sign of high texture content and a low MAD value is a sign of low texture content. Three different window sizes are used. As expressed in (8.2), a 7×7 window is used if the MAD of the center pixel is high, and a 25×25 window is used if the MAD is very low.

$$\text{MAD}(\mathbf{c}) = \frac{1}{48} \times \sum_{\mathbf{q} \in N_c} |I_L(\mathbf{q}) - I_L(\mathbf{c})| \qquad (8.1)$$

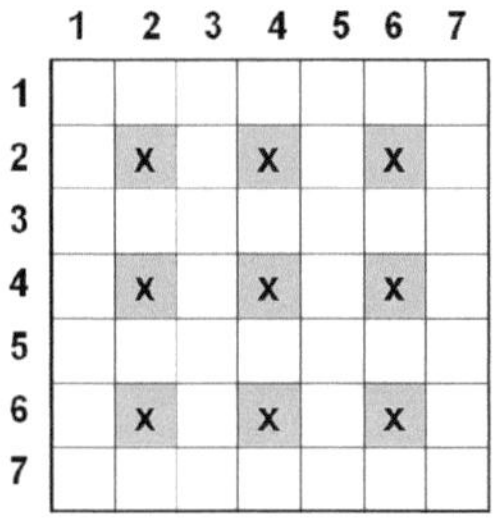

Fig. 8.1 Nine selected pixels in a block for BW-SAD calculation. Forty-nine pixels in a block are searched in parallel in hardware

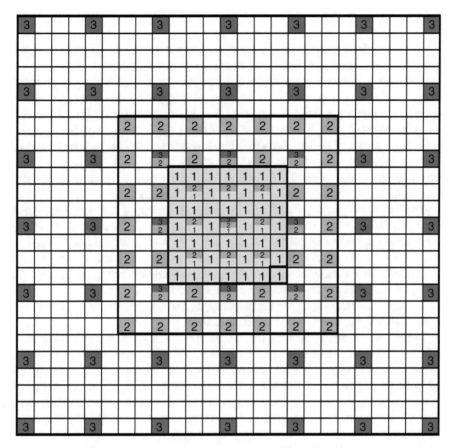

Fig. 8.2 Forty-nine selected pixels of adaptive windows (*yellow* (1): 7 × 7, *green* (2): 13 × 13, and *blue* (3): 25 × 25)

$$
\text{Window Size} = \begin{cases} 7 \times 7 & \text{if MAD}(\mathbf{c}) > \text{tr}_{7 \times 7} \\ 13 \times 13 & \text{else if MAD}(\mathbf{c}) > \text{tr}_{13 \times 13} \\ 25 \times 25 & \text{otherwise} \end{cases} \tag{8.2}
$$

As a general rule, increasing the window size increases the algorithm and hardware complexity [18]. As shown in Fig. 8.2, in our proposed algorithm, in order to provide constant hardware complexity over the three different window sizes, 49 neighbors are constantly sampled for different window sizes. "1," "2," and "3" indicate the 49 pixels used for the different window sizes 7 × 7, 13 × 13, and 25 × 25, respectively. If the sampling of 49 pixels in a window is not applied and all the pixels in a window are used during the matching process, an improvement in the disparity estimation quality can be obtained. The overhead of computational complexity for this high-complexity case and the degradation of the DE quality due to sampling are presented in Sect. 8.1.4.

8.1.1.2 Disparity Voting

A hybrid solution involving the Binary Window SAD and Census cost computation methods is presented to benefit from their combined advantages. The SAD is one of the most commonly used similarity metrics. The use of BW-SAD provides better results than using the SAD when there is disparity discontinuity since it combines information about the shape of the object with the SAD [18]. However, the computational complexity of the BW-SAD is high, thus result of this metric is provided for nine of the 49 pixels in a block and they are linearly interpolated to find the BW-SAD values for the remaining 40 pixels in a block. The selected nine pixels for the computation of BW-SAD are shown in Fig. 8.1. The low-complexity Census metric is computed for all of the 49 pixels of a block.

The formula expressing the BW-SAD for a pixel $\mathbf{p} = (x, y)$ is shown in (8.3) and (8.4). The BW-SAD is calculated over all pixels $\mathbf{q}$ of a neighborhood N_p, where the notation d is used to denote the disparity. The Binary Window, $\mathbf{w}$, is used to accumulate absolute differences of the pixels, if they have an intensity value which is similar to the intensity value of the center of the window. The multiplication with $\mathbf{w}$ in (8.4) does not cause significant computational load for the hardware since it is implemented as reset signal for the resulting absolute differences (AD). In the rest of the chapter, the term *"Shape"* is indicated by $\mathbf{w}$.

Depending on the texture of the image, the Census and the BW-SAD have different strengths and sensibility for the disparity calculation. To this purpose, a hybrid selection method is used to combine them. As shown in (8.5) and (8.6), an adaptive penalty (ap) that depends on the texture observed in the image is applied to the Hamming differences. Subsequently, the disparity with the minimum *Hybrid Cost* (HC) is selected as the disparity of a searched pixel. 2's order penalty values are used to turn the multiplication operation into a shift operation. If there is a texture on the block, the BW-SAD difference between the candidate disparities needs to be more convincing to change the decision of Census, thus a higher penalty value is applied. If there is no texture on the block, a small penalty value is applied since the BW-SAD metric is more reliable than the decision of Census.

$$\mathbf{w} = \begin{cases} 0 & \text{if } |I_L(\mathbf{q}) - I_L(\mathbf{p})| > \text{threshold}_w, \mathbf{q} \in N_p \\ 1 & \text{else} \end{cases} \tag{8.3}$$

$$\text{BW-SAD}(\mathbf{p}, d) = \sum_{\mathbf{q} \in N_p} |I_L(\mathbf{q}) - I_R(\mathbf{q} - d)| \cdot \mathbf{w} \tag{8.4}$$

$$\text{HC}(\mathbf{p}, d) = \text{BW-SAD}(\mathbf{p}, d) + \text{hamming}(\mathbf{p}, d) \times \text{ap} \tag{8.5}$$

$$\text{ap} = \begin{cases} \text{ap}_{7\times7} & \text{if} & \text{window size} = 7 \times 7 \\ \text{ap}_{13\times13} & \text{else if} & \text{window size} = 13 \times 13 \\ \text{ap}_{25\times25} & \text{else if} & \text{window size} = 25 \times 25 \end{cases} \tag{8.6}$$

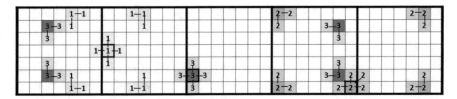

Fig. 8.3 Examples for selecting 17 contributing pixels for 7×7, 13×13, and 25×25 window sizes during the disparity refinement process (*yellow* (1): 7×7, *green* (2): 13×13, and *blue* (3): 25×25)

8.1.1.3   Disparity Refinement

The proposed disparity refinement (DR) process assumes that neighboring pixels within the same *Shape* need to have an identical disparity value, since they may belong to one unique object. In order to remove the faulty computations, the most frequent disparity value within the *Shape* is used.

As shown in Fig. 8.3, since the proposed hardware processes seven rows in parallel during the search process of a block, the DR process only takes the disparity of pixels in the processed seven rows. The DR process of each pixel is complemented with the disparities of 16 neighbor pixels and its own disparity value. Finally, the most frequent disparity in the selected 17 contributors is replaced with the disparity of that processed pixel.

The selection of these 17 contributors proceeds as follows. The disparity of the processed pixel and the disparity of its four adjacent pixels always contribute to the selection of the most frequent disparity. Four farthest possible *Shape* locations are pre-computed as a mask. If these locations are activated by *Shape*, the disparity values of these corner locations and their two adjacent pixels also contribute. Therefore, at most 17 and at least five disparities contribute to the refinement process of each pixel.

In Fig. 8.3, examples of the selection of contributing pixel locations are shown for three different window sizes. Considering the proposed contributor selection scheme, the pixels in the same row with the same window size have identical masks. The masks for the seven rows of a block and three window sizes are different. Therefore, 21 different masks are applied in the refinement process. These masks turn out to simple wiring in hardware.

Median filtering of the selected 17 contributors provides negligible improvement on the DR quality, but it requires high-complexity sorting scheme. The highest frequency selection is used for the refinement process since it can be implemented in hardware with low-complexity equality comparators and accumulators. The maximum number of contributors is fixed to 17 which provides an efficient trade-off between hardware complexity and the disparity estimation quality.

8.1.2   Hardware Implementation of Proposed Binocular AWDE Algorithm

The efficient hardware implementation of the proposed hardware-oriented binocular AWDE algorithm is presented in this section. The proposed hardware architecture of the AWDE algorithm enables handling 60 fps on a Virtex-5 FPGA at a 1024×768 XGA video resolution for a 128 pixel disparity. The proposed hardware provides dynamic and static configurability to have satisfactory disparity estimation quality for the images with different contents. It provides dynamic reconfigurability to switch between window sizes of 7×7, 13×13, and 25×25 pixels in runtime to adapt to the texture of the image. In addition, it provides static configurability to allow users to change the disparity range, the strengths of Census and BW-SAD in *HC* computation, the closest and furthest expected distances, the used color domain (Y, Cb, or Cr), etc.

8.1.2.1   Overview

The top-level block diagram of the proposed reconfigurable disparity estimation hardware and the required embedded system components for the realization of the full system are shown in Fig. 8.4. The details of main real-time video processing

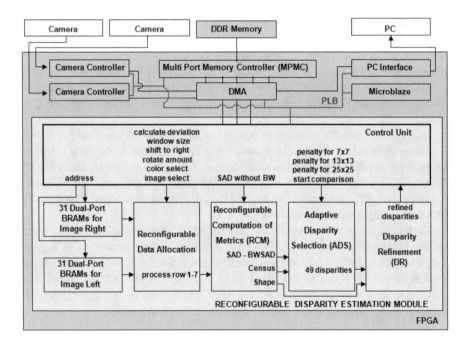

Fig. 8.4 Top-level block diagram of the disparity estimation module

hardware core of binocular disparity estimation are presented in this section. The proposed reconfigurable disparity map estimation module involves five sub-modules and 62 dual port BRAMs. These five sub-modules are the control unit, reconfigurable data allocation, reconfigurable computation of metrics (RCM), adaptive disparity selection (ADS), and disparity refinement. 31 of the 62 BRAMs are used to store 31 consecutive rows of the right image, and the remaining 31 BRAMs are used to store 31 rows of the left image. The dual port feature of the BRAMs is exploited to replace processed pixels with the new required pixels during the search process. The proposed hardware is designed to find the disparity of the pixels in the left image by searching candidates in the right image. The pixels of the right image are not searched in the left image, and thus cross-check of the DE is not applied.

The external memory bandwidth is an important limitation for disparity estimation of HR images. For example, the disparity estimation of a 1024×768 resolution stereo video at 60 fps requires 540 MB/s memory bandwidth considering loading and reading each image one time. The ZBT SRAM and DDR2 memories that are mounted on FPGA prototyping boards can typically reach approximately 1 GB/s and 5 GB/s, respectively. However, an algorithm or hardware implementation that requires multiple reads of a pixel from an external memory can easily exceed these bandwidth limitations. Using multiple stereo cameras in future targets or combining different applications in one system may bring external memory bandwidth challenges. The hardware in [6] needs to access external memory at least five times for each pixel. The hardware presented in [9] requires external memory accesses at least seven times for each pixel assuming that the entire data allocation scheme is explained. Our proposed memory organization and data allocation scheme require reading each pixel only one time from the external memory during the search process. In addition, it can be adapted to receive stream input and provide stream output without using external memory when the number of input buffer BRAMs is increased from 62 to 78.

The system timing diagram of the AWDE is presented in Fig. 8.5. The disparity refinement process is not applied to the pixels that belong to the two blocks at the right and left edges of the left image. For the graphical visualization of the reconfigurable disparity computation process together with the disparity refinement process, the timing diagram is started from the process of the sixth block of the left image. As presented in Fig. 8.5, efficient pipelining is applied between the disparity refinement and disparity selection processes. Therefore, the disparity refinement process does not affect the overall system throughput but only increases the latency. The system is able to process 49 pixels every 197 clock cycles for a 128 pixel disparity search range. Important timings during the processes are also presented with dashed lines along with their explanations.

8.1.2.2 Data Allocation and Disparity Voting

The block diagram of the reconfigurable data allocation module is shown in Fig. 8.6. The data allocation module reads pixels from BRAMs, and depending on

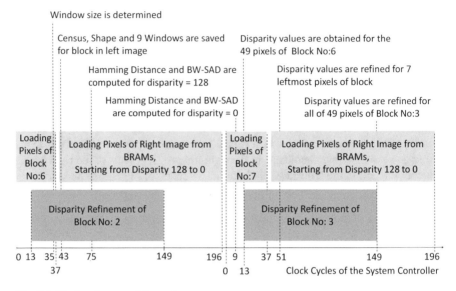

Fig. 8.5 Timing diagram of the system

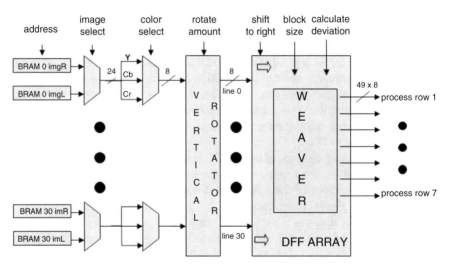

Fig. 8.6 Block diagram of the reconfigurable data allocation module

the processed rows, it rotates the rows using the vertical rotator to maintain the consecutive order. This process is controlled by the control unit through the *rotate amount* signal. The search process starts with reading the 31×31 size window of the searched block from the BRAMs of the left image. Therefore, the control unit sends the *image select* signal to the multiplexers that are shown in Fig. 8.6 to select the BRAMs of the left image. Moreover, the *color select* signal provides static configurability to select one of the pixel's components (Y, Cb, or Cr) during the search process. This user-triggered selection is useful if the Y components of the pixels are not well distributed on the histogram of the captured images. While the windows of the searched block are loaded to the D flip-flop (DFF) array, the RCM computes and stores the 49 Census transforms, 49 *Shapes*, and 9 windows pertaining to the pixels in the block for the computation of BW-SAD.

The Census transforms and windows of the candidate pixels in the right image are also needed for the matching process. After loading the pixels for the computation of metrics for the 7×7 block, the control unit selects the pixels in the right image by changing the *image select* signal, and starts to read the pixels in the right image from the highest level of disparity by sending the address signals of the candidate pixels to the BRAMs.

The disparity range can be configured by the user depending on the expected distance to the objects. Configuring the hardware for a low disparity range increases the hardware speed. In contrast, a high disparity range allows the user to find the depth of close objects. The architecture proposed in [9] is not able to provide this configurability since it is designed to search 80 disparity candidates in parallel, instead of providing parallelization to search multiple pixels in the left image. Therefore, a fixed amount of disparities is searched in [9], and changing the disparity range requires a redesign of their hardware.

The detailed block diagram of the DFF Array and the Weaver is shown in Fig. 8.7. They are the units of the system that provide the configurability of the adaptive window size. As a terminology, the term *weaving* is used to denote selecting 49 contributor pixels in different window sizes 7×7, 13×13 and 25×25 by skipping 1, 2 and 4 pixels respectively. Seven rows and one column are processed in parallel by the Weaver, and the processed pixels flow inside the DFF Array from the left to the right. Additionally, the weaving process is applied to the location (15, 8) of the DFF Array at the beginning of the search process only, to select the window size by computing the deviation of the center of the block from its neighbors for 7×7 and 13×13 windows.

The DFF Array is a 31×25 array of 8-bit registers shown in Fig. 8.7. The DFF Array has 25 columns since it always takes the inputs of the largest window size, i.e., 25×25, and it has $12 + 12 + 7 = 31$ rows to process seven rows in parallel. While the pixels are shifting to the right, the Weaver is able to select the 49 components of the different window sizes from the DFF Array with a simple wiring and multiplexing architecture. Some of the contributor pixels of the windows for different window sizes are shown in Fig. 8.7 in different colors. The Weaver and DFF Array are controlled by the control unit through the *calculate deviation*, *window size*, and *shift to right* signals. The Weaver sends seven windows to be processed by RCM as *process row 1* to *process row 7*, and each *process row* consists of 49 selected pixels.

A large window size normally involves high amounts of pixels and thus requires more hardware resources and computational cost to support the matching process. By using the proposed weaving architecture, even if the window size is changed, the windows only consist of 49 selected pixels. Therefore, the proposed hardware

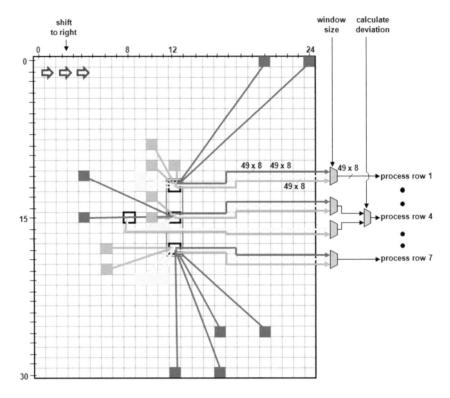

Fig. 8.7 DFF Array and the Weaver (*yellow*: 7 × 7, *green*: 13 × 13, and *blue*: 25 × 25)

architecture is able to reach the largest window size (25 × 25) among the hardware architectures implemented for DE [6, 9, 10, 12, 13, 16, 21]. The adaptability of the window size between the small and large window sizes provides high-quality disparity estimation results for HR images.

During the weaving process of the 49 pixels in the block and the candidate pixels in the right image, the RCM computes the Census and *Shape* of these pixels in a pipeline architecture. The block diagram of the RCM is shown in Fig. 8.8. The process for each block starts by computing and storing the Census and *Shape* results for the 7 × 7 block. In Fig. 8.8, the registers are named as $Shape_{row\_column}$ and $Census_{row\_column}$. Since the BW-SAD is only applied for 9 of the 49 pixels, the BW-SAD computation sub-modules are only implemented in *process rows* 2, 4, and 6.

The BW-SAD sub-module in Fig. 8.8 takes the *Shape*, registered window of the pixel in a block and the candidate window of the searched pixel as inputs, and provides the BW-SAD result as an output. The computation of the Hamming distance requires significantly less hardware area than the BW-SAD. Therefore, the Hamming computation is used for all of the 49 pixels in a block.

As shown in Fig. 8.8, when a new candidate Census for the *process row* 1 is computed by the Census sub-module of the RCM, its Hamming distance with the

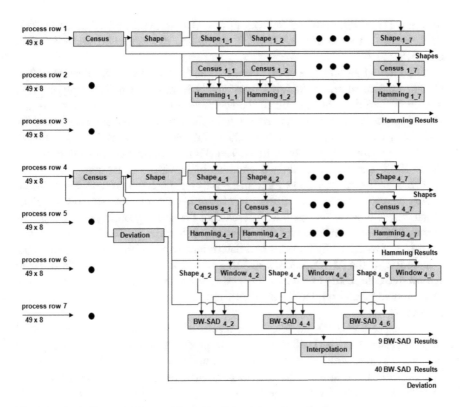

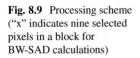

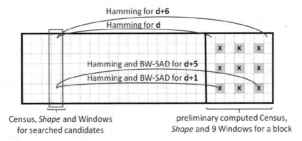

Fig. 8.8 Block diagram of the reconfigurable computation of metrics

Fig. 8.9 Processing scheme ("x" indicates nine selected pixels in a block for BW-SAD calculations)

preliminary computed seven $\text{Census}_{1\_[1:7]}$ of the block is computed by the seven Hamming sub-modules. The seven resulting Hamming Results of the *process row 1* are passed to the ADS module. Since this process also progresses in parallel for seven *process rows*, the proposed hardware is able to compute the Hamming distances of 49 pixels in a block in parallel. This parallel processing scheme is presented in Fig. 8.9. While the proposed architecture computes the Hamming distance for the leftmost pixels of the block, the Hamming for disparity d, rightmost pixels of the block computes their Hamming for disparity $d + 6$. Therefore, the resulting Hamming costs are delayed in the ADS to synchronize the costs. This delay is also an issue of the BW-SAD results and they are also synchronized in the ADS.

The internal architecture of the Census transform involves 48 subtractors. The Census module subtracts the intensity of center from the 48 neighboring pixels in a window, and uses the sign bit of the subtraction to define 48-bit Census result. The *Shape* computation module reuses the subtraction results of Census module. The *Shape* module takes the absolute values of the subtraction results and compares the absolute values with the threshold$_w$. The Hamming computation module applies 48-bit XOR operation and counts the number of 1s with an adder tree.

The deviation module shown in Fig. 8.8 only exists on the *process row 4* since it is only needed for the center of the 7×7 block to determine the window size. The module accumulates the absolute difference of the 48 neighboring pixels from the center. The control unit receives the deviation result of the 7×7 and 13×13 window sizes in consecutive clock cycles and determines window size. The mathematical calculation of the MAD requires dividing the total deviation by 48. In order to remove the complexity of the division hardware, the thresholds $tr_{7 \times 7}$ and $tr_{13 \times 13}$ are re-computed by multiplying them with 48 and compared with the resulting absolute deviations.

The use of BW-SAD provides better results than using the SAD in presence of disparity discontinuities [18]. However, if the processed image involves a significant amount of texture without much depth discontinuity, using the regular SAD provides better results. Especially for the 7×7 window size, using SAD instead of BW-SAD provides better visual results since it is the sign of significantly textured region. In order to take advantage of this property, dynamic configurability is provided to change the BW-SAD computation metric to the SAD computation for a 7×7 window. The SAD module computes the ADs and the result of ADs is stored in registers prior to accumulation. An active-low reset signal is used at the register of the AD to make its result 0, when the architecture is configured for the BW-SAD, and the respective *Shape* of the pixel in the block is 0. Otherwise, the AD register takes its actual value.

The ADS module which is shown in Fig. 8.4 receives the Hamming results and the BW-SAD results from the RCM block and determines the disparity of the searched pixels. Since the BW-SAD results are computed for 9 of the 49 pixels, the RCM linearly interpolates these nine values to find the estimated BW-SAD results of the remaining 40 pixels in the block. Due to an efficient positioning of the nine pixels in a block, the linear interpolation requires a division by 2 and 4, which are implemented as shift operations.

The ADS module shifts the Hamming results of the candidate pixels depending on the 2's order adaptive penalty for the multiplication process as shown in formula (8.5). The ADS module adds the resulting Hamming penalty on the BW-SADs to compute *Hybrid Costs*. Forty-nine comparators are used to select the 49 disparity results that point minimum *Hybrid Costs*.

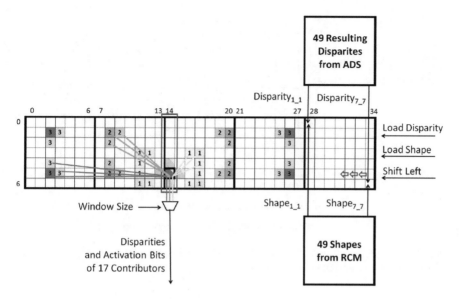

Fig. 8.10 Disparity refinement-array of the disparity refinement module (*yellow* (1): 7 × 7, *green* (2): 13 × 13, and *blue* (3): 25 × 25)

8.1.2.3 Disparity Refinement

The DR module receives the 49 disparity results from the ADS and the *Shapes* of the 49 pixels of a block from the RCM and determines the final refined disparity values. As presented in Fig. 8.10, after the ADS module has computed 49 disparity values in parallel, it loads this data in to the DFF Array of the DR module (DR-Array). The DR-Array has a size of five blocks for the refinement process. The control unit enables the DFFs by using the *Load Disparity* signal when the 49 disparity outputs of ADS module are ready for the refinement process. In each cell of the DR-Array, the respective *Shape* of a pixel is loaded from the RCM using the *Load Shape* signal. DR-Array is designed to shift the disparity and *Shape* values from right to left to allocate data for the refinement processes.

The DR hardware contains a highest frequency selection (HFS) module that consists of seven identical processing elements (DR-PE). As presented in Fig. 8.10, DR-PEs are positioned to refine seven disparities in the 15th column of the DR-Array in parallel while the disparity and *Shape* values shift through the DR-Array. The hardware architecture of a single DR-PE is presented in Fig. 8.11. The location of a single DR-PE is indicated in the sixth row of the DR-Array with a bold square.

In Fig. 8.10, while 17 disparity values are selected by the multiplexers, the *Shape* information corresponding to the four corners is also selected from the 49-bit *Shape* information of the processed pixel. The selected 4-bits inform the DR-PE which of

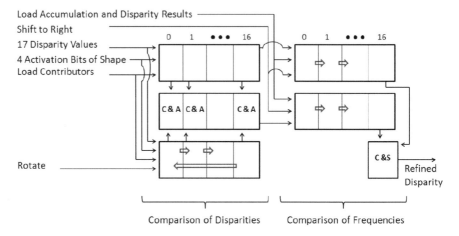

Fig. 8.11 Processing element of the disparity refinement module. The highest frequency selection module includes seven of these DR-PE elements

these 12 disparity values on the corners will be used while computing the highest frequency disparity. These 4 bits of the *Shape* are called *activation bits* in Fig. 8.11. Each *activation bit* activates itself together with its two adjacent disparities. Since the center disparity and its four neighbors are always activated, the 17-bit activation information is loaded to the DR-PE together with the respective disparities.

As presented in Fig. 8.11, the DR-PE hardware consists of two parts: Comparison of Disparities and Comparison of Frequencies. In the Comparison of Disparities part, the 17-bit activation information and the 17 disparities are stored into two DFF Arrays. One of these DFF Arrays is used as a reference and the other one rotates to compare each disparity with the 16 other disparities. During the rotation process, 17 Compare and Accumulate (C&A) sub-modules compare the disparities in parallel. If the compared disparities are identical and both of them are activated, the values of the accumulators are increased by one. After 17 clock cycles, the values in the accumulators and their respective disparities are loaded into the DFF Array in the Comparison of Frequencies part of the DR-PE. In the pipeline architecture, at the same time, the control unit shifts the DR-Array to the left by one to load new 17 contributors to the DR-PE. The Compare and Select (C&S) sub-module compares the values of the accumulators to find the highest value in the accumulators, and selects the disparity with the highest frequency as the refined disparity. Since the DR process works in parallel with the other hardware modules of AWDE, it does not affect the throughput of the DE system if the disparity range is configured as more than 70.

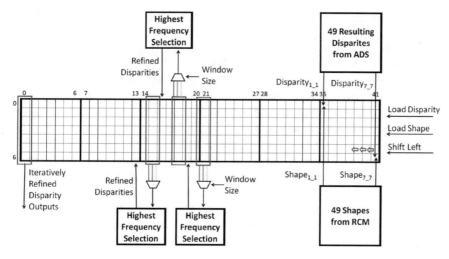

Fig. 8.12 DR-Array of the iterative disparity refinement module (*yellow line*: 7 × 17 candidates for 7 × 7 window, *green line*: candidates for 13 × 13, and *blue line*: candidates for 25 × 25)

8.1.3 Iterative Refinement for the Enhanced AWDE Implementation

The intuition behind the proposed Iterative Refinement process of the IR-AWDE algorithm is identical to the DR process presented in Sect. 8.1.1.3: neighboring pixels within the same *Shape* need to have an identical disparity value, since they may belong to one unique object. Using the refinement process multiple times removes noisy computations more efficiently, and increases the disparity estimation quality.

The iterative refinement hardware is presented in Fig. 8.12 which consists of an improved version of the DR hardware presented in Fig. 8.10. The proposed Iterative Refinement process utilizes three concatenated highest frequency selection modules. Each HFS module includes seven identical DR-PEs, one of which is presented in Fig. 8.11. All DR-PEs receive 17 selected disparities from their own multiplexer. The DR-Array in Fig. 8.10 includes DFFs to keep record of the computed disparities for five blocks. Instead, for the IR, the size of the DFF-Array is increased to six blocks since the disparities need to be pipelined for longer duration. Moreover, the DR hardware presented in Fig. 8.10 provides the most frequent disparities as an output as the refined disparities. Instead, the HFS modules for the IR hardware write back the refined disparities on DR-Array. Writing back the most frequent disparities into the DR-Array provides an iterative refinement of the estimated disparities. Since the disparity results shift inside the DR-Array, refined disparities are overwritten two pixels left of the consecutive pixel location. For example, as presented in Fig. 8.12, while the HFS module refines the disparities of the seven pixels in column 21 of the DR-Array, the DR-Array shifts the disparity values two times. Therefore, the computed seven highest frequency disparities in the column 19 of the DR-Array are overwritten.

In addition to removing noisy computations, IR provides efficient results in assigning disparities of occluded regions. While searching pixels from the left image inside the right image, occluded regions appear on the left side of objects [19]. Consequently, wrong computations due to occlusion appear on the left sides of the objects in the image, which should be replaced by the correct disparities that are assigned to the left adjacent pixels of the occluded ones. The proposed iterative refinement process scans the estimated disparities from left to right. In addition, HFS modules receive updated disparities from their left since they are already overwritten by the refined ones. Therefore, this process iteratively spreads the correct disparities to the occluded regions while considering the object boundaries with the *Shape* information. While disparities shift inside the DR-Array, the leftmost disparities in the column 0 of the DR-Array are provided as the refined disparity value outputs of the IR Module.

8.1.4 Implementation Results

The reconfigurable hardware architecture of the proposed AWDE algorithm is implemented using Verilog HDL, and verified using Modelsim 6.6c. The Verilog RTL models are mapped to a Virtex-5 XCUVP-110T FPGA comprising 69k look-up tables (LUT), 69k DFFs, and 144 Block RAMs (BRAM). The proposed hardware consumes 59% of the LUTs, 51% of the DFF resources, and 42% of the BRAM resources of the Virtex-5 FPGA. The proposed hardware operates at 190 MHz after place and route and computes the disparities of 49 pixels in 197 clock cycles for 128 pixel disparity range. Therefore, it can process 60 fps at a 1024×768 XGA video resolution.

The AWDE-IR is implemented to further improve the disparity estimation quality of AWDE using an efficient iterative refinement step. The hardware implementation of AWDE-IR is mapped to a same FPGA and verified using Modelsim 6.6c. The proposed AWDE-IR hardware consumes 70% of the LUTs, 63% of the DFF resources, and 42% of the BRAM resources of the Virtex-5 FPGA. It can work at same speed performance due to the pipeline structure of the refinement process.

The parameters of the AWDE algorithm are shown in Table 8.1. Parameters are selected by sweeping to obtain high-quality DE of HR images considering different features pertaining to the image content.

Tables 8.2 and 8.3 compare the disparity estimation performance and hardware implementation results of the AWDE architecture with other existing hardware

Table 8.1 Parameters of the AWDE

$tr_{7\times7}$	$tr_{13\times13}$	$ap_{7\times7}$	$ap_{13\times13}$	$ap_{25\times25}$	threshold$_w$
5	2	32	16	4	8

Table 8.2 Disparity estimation performance comparisons

	Tsukuba (288 × 384)	Venus (383 × 434)	Aloe (1110 × 1282)	Art (1110 × 1390)	Clothes (1110 × 1300)
Chang [6]	4.15	0.56	3.75	12.80	2.97
Ttofis [21]	13.21	4.56	8.88	32.18	7.67
Greisen [10]	12.42	4.14	8.65	23.46	5.30
Georgoulas [9]	12.38	15.20	6.97	23.75	9.15
Census7	26.05	30.80	20.36	45.39	21.80
Census13	18.19	18.83	11.21	31.65	9.36
Census25	15.94	15.38	10.41	29.66	7.16
BWSAD7	12.19	19.45	8.31	34.03	13.33
BWSAD13	11.23	15.16	7.13	28.57	9.27
BWSAD25	10.43	11.12	6.74	24.74	6.28
FWDE7	9.53	12.59	5.38	20.87	5.39
FWDE13	7.90	6.82	4.81	16.97	3.16
FWDE25	8.03	5.66	5.16	18.12	3.87
AWDE	7.64	5.33	4.94	16.33	2.89
AWDE-HC	7.47	4.73	4.92	16.17	2.95
AWDE-IR	6.53	5.01	4.30	14.47	2.94

Error rates (%) are provided compared to DE ground truths of the benchmark pictures

Table 8.3 Hardware performance comparison

Hardware	Technology	Image resolution	DFF consumption	LUT consumption	Disparity range	fps	Clock speed (MHz)
Chang [6]	ASIC-90 nm	352 × 288	562k gates		64	42	95
Ttofis [21]	Virtex-5	1280 × 1024	31k	47k	120	50	100
Greis. [10]	Stratix-III	1920 × 1080	26k	54k	256	30	130
Georg. [9]	Stratix-IV	800 × 600	15k	146k	80	550	511
		1024 × 768			128	60	
AWDE	Virtex-5	640 × 480	35k	40k	64	221	190
		352 × 288			64	670	
		1024 × 768			128	60	
AWDE-IR	Virtex-5	640 × 480	43k	48k	64	221	190
		352 × 288			64	670	

implementations that targets HR [9, 10, 21] and currently the highest quality DE hardware that targets LR [6]. These papers do not provide the disparity estimation quality results for the HR benchmarks of the Middlebury dataset [19]. Thus, we implemented [6, 9, 10] in software, and the software implementation of [21] is obtained from its authors. The DE results for the Census and the BW-SAD metrics for different window sizes are also presented in Table 8.2. The comparisons of the resulting disparities with the ground truths are done as prescribed by the Middlebury evaluation module. If the estimated disparity value is not within a one range of the ground truth, the disparity estimation of the respective pixel is considered erroneous. 18 pixels located on the borders are neglected in the evaluation of LR benchmarks Tsukuba and Venus, and a disparity range of 30 is applied for all algorithms. Thirty pixels located on the borders are neglected in the evaluation of HR benchmarks Aloe, Art, and Clothes, and a disparity range of 120 is applied for all algorithms.

The Census and BW-SAD results that are shown in Table 8.2 are provided by sampling 49 pixels in a window. FW-DE indicates the combination of BW-SAD and Census for a fixed window size. The numbers terminating the name of the algorithms indicate the fixed window sizes of these algorithms.

Although the Census and the BW-SAD algorithms do not individually provide very efficient results, the combination of these algorithms into the FW-DE provides an efficient hybrid solution as presented in Table 8.2. For example, if a 7×7 window size and Census method are exclusively used for DE on the HR benchmark Art, 45.39% erroneous DE computation is observed from the result of Census7. Exclusively using a 7×7 window size and BW-SAD method for the same image yields 34.03% erroneous computation. However, if only a 7×7 window size is used combining the Census and BW-SAD methods, 20.87% erroneous computation is observed as presented in the result of FW-DE7. 20.87% erroneous computation is significantly smaller than 45.39% and 34.03%, which justifies the importance of combining the Census and BW-SAD into a hybrid solution. For the same image, using the FW-DE13 and FW-DE25 algorithms yields 16.97% and 18.12% erroneous DE computations, respectively. Combining the FW-DE7, FW-DE13, and FW-DE25 into a reconfigurable hardware with an adaptive window size feature further improves the algorithm results as demonstrated from the results of AWDE. AWDE provides 16.33% erroneous computation for the same image which is smaller than 20.87%, 16.97%, and 18.12%, thus numerically emphasizing the importance of adaptive window size selection. The algorithmic performance of AWDE, 16.33%, is considerably better than the DE performance results of HR DE hardware implementations [10, 21] and [9] that provide 32.18%, 23.46%, and 23.75% erroneous computations, respectively, for the same image.

If the sampling of 49 pixels in a window is not applied and all the pixels in a window are used during the matching process, the complexity of the AWDE algorithm increases by 12 times. The result of the high-complexity version of the AWDE algorithm (AWDE-HC) is also provided in Table 8.2 for comparison. The AWDE-HC provides almost the same quality results as the AWDE. Considering the hardware overhead of AWDE-HC, the low-complexity version of the algorithm, AWDE, is selected for hardware implementation, and its efficient reconfigurable hardware is presented.

Improving the results of AWDE is possible using the low-complexity iterative refinement step as indicated from the results of AWDE-IR. AWDE-IR efficiently removes a significant amount of noisy computations by iteratively replacing the disparity estimations with the most frequent neighboring ones as can be observed from the results of Tsukuba, Venus, Aloe, and Art. Moreover, IR does not require significant amount of additional computational complexity. Therefore, AWDE-IR is implemented in hardware for the further improvement of the disparity estimation quality.

The algorithm presented in [6] uses the Census algorithm with the cost aggregation method, and provides the best results for both LR and HR stereo images except the HR benchmark Clothes. As shown in Table 8.3, due to the high-complexity of cost aggregation, it only reaches 42 fps for CIF images, thereby consuming a large amount of hardware resource. If the performance of [6] is scaled to 1024 × 768 for a disparity range of 128, less than 3 fps can be achieved.

None of the compared algorithms that have a real-time HR hardware implementation [9, 10, 21] is able to exceed the DE quality of AWDE and AWDE-IR for HR images. The overall best results following the results of AWDE and AWDE-IR are obtained from [10]. The hardware presented in [10] consumes 20% of the 270k Adaptive LUT (ALUT) resources of a Stratix-III FPGA. It provides high disparity range due to its hierarchical structure. However, this structure easily causes faulty computations when the disparity selection finds wrong matches in low resolution.

The hardware implementation of [9] provides the highest speed performance in our comparison. However, this hardware applies 480 SAD computations for a 7 × 7 window in parallel. The hardware presented in [9] consumes 60% of the 244k ALUT resources of a Stratix-IV FPGA. In our hardware implementation we only use nine SAD computations in parallel for the same size window and this module consumes 16% of the resources of a Virtex-5 FPGA on its own. Therefore, the hardware proposed in [9] may not fit into three Virtex-5 FPGAs.

The visual results of the AWDE and AWDE-IR algorithms for the HR benchmarks Clothes, Art, and Aloe are shown in Figs. 8.13, 8.14, and 8.15, respectively. The disparity map result of the AWDE algorithm for the 1024 × 768 resolution pictures taken by our stereo camera system is shown in Fig. 8.16. The proposed binocular disparity estimation hardware architectures provide both quantitative and visual satisfactory results and they reach real-time for HR.

8.2   Trinocular Adaptive Window Size Disparity Estimation Algorithm and Its Hardware Implementation

This section presents a hardware-oriented trinocular adaptive window size disparity estimation (T-AWDE) algorithm and the first real-time trinocular disparity estimation (DE) hardware that targets high-resolution images with high-quality disparity results [3]. The proposed trinocular DE hardware is the enhanced version of the binocular AWDE implementation that is presented in Sect. 8.1. The T-AWDE hardware generates a very high-quality depth map by merging two depth maps obtained from the center-left and center-right camera pairs. The T-AWDE hardware enhances disparity results by applying a double checking scheme which solves most of the occlusion problems existing in the AWDE implementation while providing

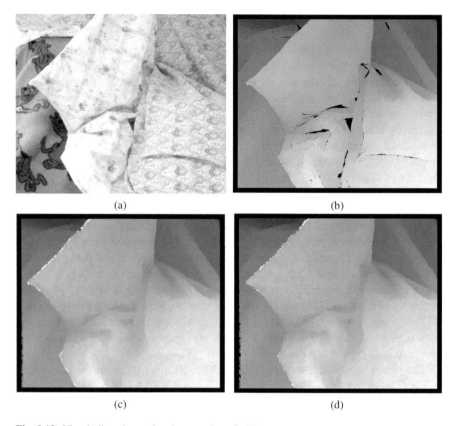

Fig. 8.13 Visual disparity estimation results of AWDE and AWDE-IR algorithms for HR benchmark Clothes. *Black regions* in the ground truths are not taken into account for the error computations as explained in Middlebury evaluation. (**a**) Left image, (**b**) ground truth, (**c**) DE result of AWDE, (**d**) DE result of AWDE-IR

correct disparity results even for objects located at left or right edge of the center image.

8.2.1   Trinocular Hardware-Oriented Adaptive Window Size Disparity Estimation Algorithm

The proposed T-AWDE algorithm is developed to support efficient parallel operations, to consume low hardware resources, and to avoid the requirement of an external memory while providing very high-quality DE results. As presented in Fig. 8.17, while processing the trinocular DE for every pixel of the center image, the candidate disparities on the right side are searched for the center-left pair, and the candidate disparities on the left side are searched for the center-right pair. Therefore, in the T-AWDE algorithm, two disparity maps are calculated for the center-left and

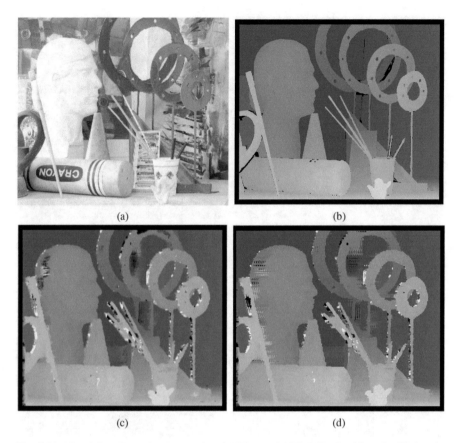

Fig. 8.14 Visual disparity estimation results of AWDE and AWDE-IR algorithms for HR bench-mark Art. *Black regions* in the ground truths are not taken into account for the error computations as explained in Middlebury evaluation. (**a**) Left image, (**b**) ground truth, (**c**) DE result of AWDE, (**d**) DE result of AWDE-IR

center-right pairs. The T-AWDE algorithm combines these two disparity maps to provide a very high-quality DE result for the center camera. The proposed T-AWDE algorithm consists of six main parts: preprocessing, window size determination, matching cost calculation, disparity selection, fusion of the disparity maps, and iterative refinement (IR).

Image rectification is one of the most essential preprocessing parts of DE. The rectification process requires internal and external calibrations to model distortions of the lenses and the mechanical misalignment of the cameras. The Open-CV calibration toolbox [5] is used for external and internal calibrations. The Caltech rectification algorithm [4] is used to horizontally align the images captured from three cameras.

The window size determination, matching cost calculation, and binocular dispar-ity selection parts of the T-AWDE algorithm for each camera pair are identical to the AWDE algorithm. As presented in Fig. 8.17, the T-AWDE searches 49 pixels of the

Fig. 8.15 Visual disparity estimation results of AWDE and AWDE-IR algorithms for HR benchmark Aloe. *Black regions* in the ground truths are not taken into account for the error computations as explained in Middlebury evaluation. (**a**) Left image, (**b**) ground truth, (**c**) DE result of AWDE, (**d**) DE result of AWDE-IR

center image in parallel in candidate disparities of the right and left images. Each of these 49 pixels is independently searched, using their own window. The selection of a large window size improves the algorithm performance in textureless regions while requiring higher computational load. However, the usage of small window sizes provides better disparity results in regions where the image has a texture. The T-AWDE dynamically changes the windows size as either 7×7, 13×13, or 25×25 pixels to adapt to different texture features of the images. It utilizes the mean absolute deviation (MAD) to measure the local texture feature of the center image, and compares the MAD values with the threshold values to adaptively determine the window size. A constant hardware complexity over the three different window sizes is provided by constantly selecting 49 contributor pixels in different

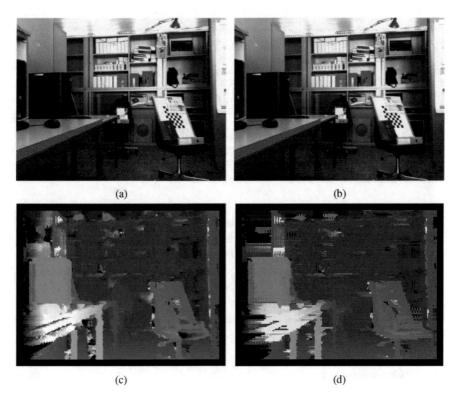

(a)   (b)

(c)   (d)

Fig. 8.16 Visual disparity estimation results of AWDE and AWDE-IR algorithms for the 1024 × 768 resolution pictures captured by the implemented stereo camera system. The ground truth for these images is not available. (**a**) Left image, (**b**) right image, (**c**) DE result of AWDE, (**d**) DE result of AWDE-IR

window sizes of 7×7, 13×13, and 25×25 pixels by skipping 1, 2, and 4 pixels, respectively. The T-AWDE adaptively utilizes the Census and binary-window sum of absolute difference (BW-SAD) matrixes as a hybrid solution. The computation of the Hybrid Cost (HC) provides high-quality results for object boundaries and adapts to different light conditions of real images. Further details about window size determination, matching cost calculation, and binocular disparity selection are presented in Sect. 8.1.

The fusion process includes two steps. In the first step, the T-AWDE compares HC values of the center-left and center-right pairs for every disparity, in order to select the one that exhibits the minimum cost as a disparity value. The winner-take-all (WTA) approach provides high-quality results especially in occluded and low-textured regions, thanks to the strengths of the AWDE algorithm and the usage of three horizontally aligned cameras. In the next step of the fusion, a *confidence* metric is computed to further improve the DE results. The expression of the *confidence* metric is presented in (8.7). Here, c_1 represents the minimum matching cost and d_1 is the corresponding disparity value of the minimum matching cost. c_2 represents the

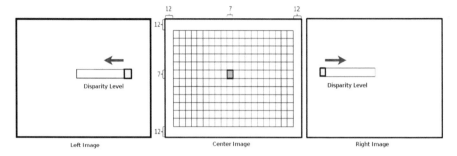

Fig. 8.17 Matching directions of the T-AWDE algorithm

second minimum matching cost and d_2 is the corresponding disparity value of the second minimum matching cost. (c_1, d_1) and (c_2, d_2) belong to the overall matching cost computation obtained from any of the stereo pairs. This metric identifies the strength of the global minima compared to local minima. Ambiguity in the selected minima is identified when two low-value costs are detected within the disparity range, while very different attached distances are computed. The *confidence* value of d_1 is then assigned as 0. The *confidence* metric is used in the disparity refinement process.

$$\text{Confidence} = \begin{cases} 0 & (c_1 - c_2) \leq (c_1/4) \text{ and } (|d_1 - d_2|) > 5 \\ 1 & \text{otherwise} \end{cases} \qquad (8.7)$$

The T-AWDE smoothens the computed disparity map using the brightness values of the neighboring pixels following the IR scheme prescribed in Sect. 8.1. The refinement process assumes that neighboring pixels with similar brightness values need to have identical disparity values, since they may belong to one unique object. In the refinement process, the disparities that are not frequently observed in the neighborhood are considered as faulty computations, and they are replaced by the most frequent disparity value that is computed in the neighborhood. This process is iteratively handled from the left to the right of the disparity image. In addition to the IR scheme presented in Sect. 8.1, the *confidence* metric is used in the IR process of T-AWDE algorithm. The unconfident disparity values are disregarded while determining the most frequent disparity value in order to prevent propagating them into the final DE result. Using the *confidence* metric during the IR process eliminates a significant amount of incorrect propagations, especially within low-textured regions.

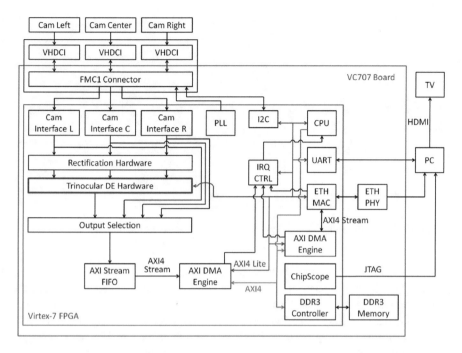

Fig. 8.18 Block diagram of the full system

8.2.2   Trinocular Adaptive Window Size Disparity Estimation Hardware

The block diagram of the full system that implements the T-AWDE algorithm is presented in Fig. 8.18. A Virtex-7 FPGA included in the VC707 Evaluation Board is used to prototype the developed hardware. All real-time video processing computations are implemented in hardware. The PC is used as a display, to control the system and to obtain camera calibration parameters. The resulting disparity images are transferred to the PC using 1 Gb raw Ethernet. A standard 2D TV is connected to the PC using HDMI to offer a better display. A MicroBlaze softcore is used to initialize cameras through I²C, to control Ethernet, and to communicate with the PC. A DDR3 memory is only used for Ethernet buffering. The details of main real-time video processing hardware core of trinocular disparity estimation are presented in this section.

The camera interface, rectification, and disparity estimation blocks are designed to avoid using the DDR3 memory; thus they receive a streaming input and provide a streaming output. Therefore, the video processing core can be easily converted to a single ASIC. Eliminating the DDR3 memory from the video processing is a benefit of the efficient and hardware-oriented algorithm that only requires local processing, the utilization of local on-chip Block RAMs (BRAMs), and the perfect

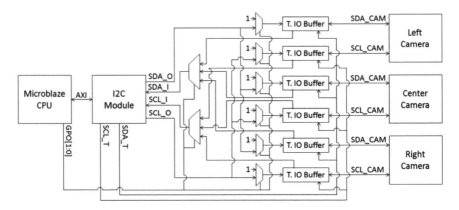

Fig. 8.19 I^{2}C multiplexing hardware (resistors and level-shifters are not drawn)

synchronization of cameras. Perfect synchronization of the cameras also always keeps fast moving objects in the same epipolar lines of the cameras to provide best DE quality for these objects. The system is able to deliver 32-bit pixel RGB+Disparity video of a center camera, or RGB video of any camera. Moreover, the system can be configured to deliver trinocular DE, center-left DE, or center-right DE results as a final disparity value to provide easy comparison of the results.

The perfect synchronization of cameras is achieved by providing a common clock source to the cameras from the FPGA, and simultaneously programming the cameras using the same I^{2}C module. Using the same clock source for the cameras is necessary to obtain an identical frame rate from the cameras. Simultaneous programming of the cameras is needed to start cameras exactly at the same time. The Xilinx I^{2}C IP and I^{2}C lines of the cameras are connected through multiplexers and tristate IO buffers as presented in Fig. 8.19. All three cameras have the same I^{2}C address. The MicroBlaze controls the select modes of the multiplexers using a 2-bit general purpose output (GPO). The multiplexing scheme allows to write the control registers of three camera sensors at the same time, to write to one of the selected cameras, or to read from one of the selected cameras. This connection method allows to concurrently write control registers of three cameras by ignoring the acknowledge signal. This transmission method does not conform the definition of I^{2}C. Nevertheless, according to our real-time tests, data is always correctly transmitted if I^{2}C is used at low frequency.

The Caltech rectification hardware is implemented to solve the lens distortions and camera misalignments as a preprocessing step of trinocular disparity estimation. The implemented rectification hardware utilizes 64 on-chip BRAMs for each camera. Each BRAM is used to buffer one row of the image. The rectification hardware processes the images of three cameras in parallel, and synchronously transfers rectified YCbCr images to the disparity estimation module.

The block diagram of the trinocular DE hardware is shown in Fig. 8.20. The disparity estimation hardware buffers the input pixel values using 39 single-port

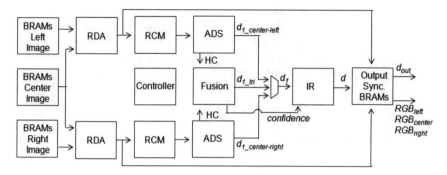

Fig. 8.20 Block diagram of trinocular DE hardware

BRAMs for each camera to realize a window-based matching scheme. The three-camera disparity estimation hardware is composed of two high-performance and high-quality binocular disparity estimators presented in Sect. 8.1. Each of these estimators includes three modules named as reconfigurable data allocation (RDA), reconfigurable computation of metrics (RCM), and adaptive disparity selection (ADS). The fusion module combines the DE outputs of the two estimators. The IR module smoothens DE computations and provides the final output of the trinocular DE. Finally, the 8-bit disparity values and the 24-bit RGB pixels of the left, center, and right cameras are buffered and synchronized using 16 BRAMs for each channel before transferring this data to the Output Selection module that is presented in Fig. 8.18. The controller generates read-addresses to the BRAMs, manages DE hardware modules to maintain their synchronous process, and interfaces with the MicroBlaze to apply user-programmable features that are provided from the GUI. In order to allow the programmability of the DE hardware, the controller includes software accessible registers. The hardware is configurable by the user who can select the maximum disparity range (maximum 255 is allowed), the disparity-start and disparity-end values to provide best DE quality at a certain distance interval, the strengths of the Census and BW-SAD metrics in the HC computation, and the resolution of depth images (maximum 1024×768 is allowed) to allow faster frame rates at lower resolution video.

The RDA module includes a vertical rotator, DFF-Array, and weaver sub-modules to arrange adaptive window sizes. The RCM computes the Census transforms, neighborhood information, BW-SAD, and Hamming costs of the 49 parallel processed pixels. The ADS module receives the Hamming results and the BW-SAD results from the RCM, computes the HC values, and determines the disparities of the 49 searched pixels pertaining to the center-left and center-right pairs.

The parallel processing scheme of the T-AWDE hardware is presented in Fig. 8.21. The search process starts by reading the windows of the 49 processed pixels from the BRAMs of the center image. The RDA and RCM compute the Census and neighborhood information of the 49 processed pixels and permanently saves these values. Subsequently, the synchronous and symmetric scanning processes of

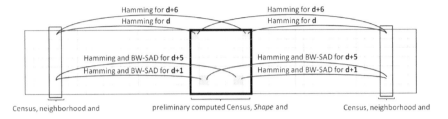

Fig. 8.21 Parallel processing scheme for two pairs

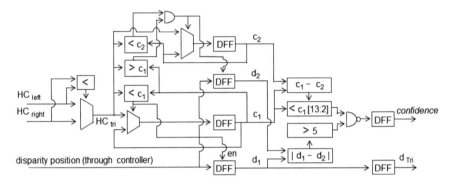

Fig. 8.22 Processing element of the fusion module (PE-F). The fusion module includes 49 PE-F elements

candidate disparities start. The controller starts to synchronously read the pixels in the right and left image from the highest level of disparity by sending the address signals of the candidate pixels to the BRAMs. The controller sends addresses to the right image BRAMs in increasing order whereas it sends addresses to the left image BRAMs in decreasing order. Although the scanning process is symmetric, due to the square shape of the processed block, the Hamming and BW-SAD values of the pairs are not synchronously computed for the same disparity. For the center-right pair, while the proposed architecture computes the Hamming values for the leftmost pixels of the block, the Hamming for disparity d, the rightmost pixels of the searched block computes their Hamming for disparity $d + 6$. Whereas, for the center-left pair, while the proposed architecture computes the Hamming values for the leftmost pixels of the block, the Hamming for disparity $d + 6$, rightmost pixels of the block computes their Hamming for disparity d. The RCM applies additional pipelining to synchronize the matching cost. This synchronization process is necessary to allow fusion module to concurrently compare HC values obtained from pairs for the same disparity values.

The fusion module compares the HC values obtained from two pairs, and applies WTA to determine the trinocular disparity values of 49 pixels in parallel. Moreover it computes *confidence* values to be used for the IR process. The processing element of fusion hardware (PE-F) is presented in Fig. 8.22. The fusion hardware includes 49

PE-F to compute 49 trinocular disparities in parallel. The PE-F compares HC values obtained from two pairs to determine the c_1, d_1, c_2, and d_2 values using comparators and multiplexers. The PE-F computes mathematical operations of the *confidence* calculation presented in (1), and transfers d_1 and its respective *confidence* value to the disparity refinement module to compute the final disparity outputs.

The IR hardware of T-AWDE implementation is different from the hardware presented in Sect. 8.1, since the T-AWDE additionally utilizes a *confidence* metric. The IR module of the T-AWDE computes most frequent disparity value in the neighborhood and replaces noisy DE computations with the most frequent disparity values. If a particular disparity value is not deemed trustable, the value is removed from the neighborhood of other pixels by using logical AND gates. Using this method, only disparity values that are identified as confident propagate to the neighboring pixels.

8.2.3   Implementation Results

The proposed real-time trinocular DE hardware is implemented using Verilog HDL, and verified using Modelsim 10.1d. The Verilog RTL models are mapped to a Virtex 7 XC7VX485T FPGA comprising 607k look-up tables (LUT), 303k DFFs, and 1030 BRAMs. The trinocular DE hardware consumes 25% of the LUTs, 11% of the DFF, and 16% of the BRAM resources of the FPGA. The proposed hardware operates at 175 MHz after place and route, and computes the disparities of 49 pixels in 198 clock cycles for a 128 pixel disparity range. Therefore, it can process 55 fps at a 768×1024 XGA video resolution. The system is functionally verified in real-time. Although a 55 fps performance is verified using Chipscope, the current display output of the system is 18 fps due to the bandwidth limitation of raw Ethernet output. The 55 fps performance of the hardware will be fully exploitable using USB3 or HDMI.

The visual results of the T-AWDE and AWDE algorithms using the Middlebury benchmark image set Bowling2 (1276×1110) are obtained from MATLAB simulations and presented in Fig. 8.23. The comparisons of the resulting disparities with the ground truth are realized as prescribed by the Middlebury evaluation module. Using the AWDE algorithm for center-left and center-right pairs yields 18.01% and 15.60% erroneous DE computations, respectively. Combining the binocular pairs into proposed trinocular DE improves the algorithm results as demonstrated in Fig. 8.23f. The proposed T-AWDE algorithm provides 9.41% erroneous computation for the same image thus numerically emphasizes the importance of proposed trinocular DE algorithm. 1024×768 resolution real-time snapshots captured by the proposed system are presented in Fig. 8.24. Models stand stable in front of the system while capturing multiple consecutive snapshots. The center-left, trinocular,

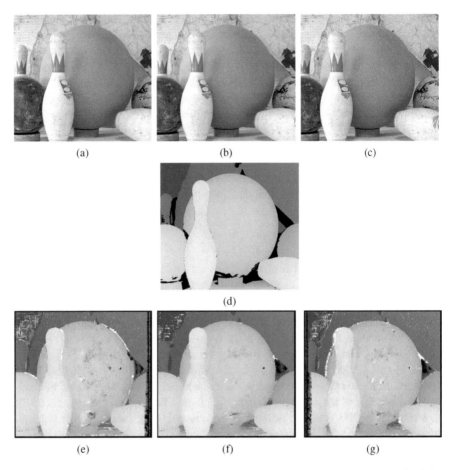

Fig. 8.23 DE results obtained by MATLAB for Middlebury benchmark image set "Bowling." (**a**) Left image, (**b**) Center image, (**c**) Right image, (**d**) Ground truth (black pixels are ignorable), (**e**) IR-AWDE for center-left (18.01%), (**f**) T-AWDE (9.41%), (**g**) IR-AWDE for center-right (15.60%)

and center-right DE results are presented in Fig. 8.24d, e, f, respectively. The T-AWDE solves a significant amount of the occlusion and incorrect estimation errors exploiting the fusion of the DE results of the two pairs. Hence, the proposed T-AWDE system delivers high-quality results and realizes the first real-time trinocular DE hardware for high resolution.

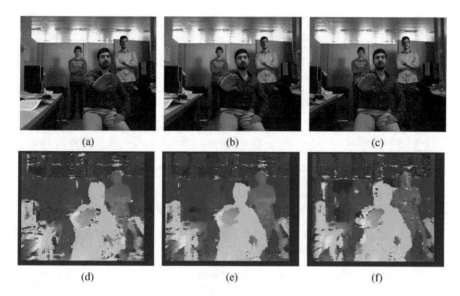

Fig. 8.24 Real-time snapshots captured by the proposed system. A ground truth for these images is not available. (**a**) Left image, (**b**) Center image, (**c**) Right image, (**d**) AWDE for center-left, (**e**) T-AWDE, (**f**) AWDE for center-right

References

1. Akin A, Baz I, Atakan B, Boybat I, Schmid A, Leblebici Y (2013) A hardware-oriented dynamically adaptive disparity estimation algorithm and its real-time hardware. In: Proceedings of the 23rd ACM international conference on Great Lakes symposium on VLSI, GLSVLSI '13. ACM, New York, NY, pp 155–160. doi: 10.1145/2483028.2483082. http://doi.acm.org/10.1145/2483028.2483082

2. Akin A, Baz I, Schmid A, Leblebici Y (2014) Dynamically adaptive real-time disparity estimation hardware using iterative refinement. Integr VLSI J 47(3):365–376. doi:http://dx.doi.org/10.1016/j.vlsi.2013.11.002. http://www.sciencedirect.com/science/article/pii/S0167926013000734. Special issue: VLSI for the new era

3. Akin A, Capoccia R, Narinx J, Schmid A, Leblebici Y (2015) Trinocular adaptive window size disparity estimation algorithm and its real-time hardware. In: International symposium on VLSI design, automation and test (VLSI-DAT). IEEE, New York, pp 1–4

4. Bouguet JY (2004) Camera calibration toolbox for matlab. [Online] Available: http://www.vision.caltech.edu/bouguetj/

5. Bradski G, Kaehler A (2008) Learning OpenCV: computer vision with the OpenCV library. O'Reilly Media, Inc., Sebastopol

6. Chang NC, Tsai TH, Hsu BH, Chen YC, Chang TS (2010) Algorithm and architecture of disparity estimation with mini-census adaptive support weight. IEEE Trans Circuits Syst Video Technol 20(6):792–805

7. Cheung TK, Woo K (2011) Human tracking in crowded environment with stereo cameras. In: 2011 17th international conference on digital signal processing (DSP). IEEE, New York, pp 1–6

8. Field M, Clarke D, Strup S, Seales WB (2009) Stereo endoscopy as a 3-d measurement tool. In: 2009 annual international conference of the IEEE engineering in medicine and biology society, EMBC 2009. IEEE, New York, pp 5748–5751

9. Georgoulas C, Andreadis I (2009) A real-time occlusion aware hardware structure for disparity map computation. In: Image analysis and processing–ICIAP 2009. Springer, New York, pp 721–730

10. Greisen P, Heinzle S, Gross M, Burg AP (2011) An FPGA-based processing pipeline for high-definition stereo video. EURASIP J Image Video Process 2011(1):1–13

11. Grosse M, Buehl J, Babovsky H, Kiessling A, Kowarschik R (2010) 3d shape measurement of macroscopic objects in digital off-axis holography using structured illumination. Opt Lett 35(8):1233–1235

12. Jin S, Cho J, Dai Pham X, Lee KM, Park SK, Kim M, Jeon JW (2010) FPGA design and implementation of a real-time stereo vision system. IEEE Trans Circuits Syst Video Technol 20(1):15–26

13. Lee SH, Sharma S (2011) Real-time disparity estimation algorithm for stereo camera systems. IEEE Trans Consum Electron 57(3):1018–1026

14. Merkle P, Morvan Y, Smolic A, Farin D, Mueller K, de With P, Wiegand T (2009) The effects of multiview depth video compression on multiview rendering. Signal Process Image Commun 24(1):73–88

15. Min D, Kim D, Yun S, Sohn K (2009) 2d/3d freeview video generation for 3DTV system. Signal Process Image Commun 24(1):31–48

16. Miyajima Y, Maruyama T (2003) A real-time stereo vision system with FPGA. In: Cheung PYK, Constantinides G (eds) Field programmable logic and application. Lecture notes in computer science, vol 2778. Springer, Berlin, pp 448–457. doi:10.1007/978-3-540-45234-8_44. http://dx.doi.org/10.1007/978-3-540-45234-8_44

17. Mori Y, Fukushima N, Yendo T, Fujii T, Tanimoto M (2009) View generation with 3d warping using depth information for FTV. Signal Process Image Commun 24(1):65–72

18. Motten A, Claesen L (2010) A binary adaptable window SoC architecture for a stereo vision based depth field processor. In: 2010 18th IEEE/IFIP VLSI system on chip conference (VLSI-SoC). IEEE, New York, pp 25–30

19. Scharstein D, Szeliski R (2002) A taxonomy and evaluation of dense two-frame stereo correspondence algorithms. Int J Comput Vis 47(1–3):7–42

20. Tombari F, Mattoccia S, Di Stefano L (2010) Stereo for robots: quantitative evaluation of efficient and low-memory dense stereo algorithms. In: 2010 11th international conference on control automation robotics & vision (ICARCV). IEEE, New York, pp 1231–1238

21. Ttofis C, Hadjitheophanous S, Georghiades A, Theocharides T (2012) Edge-directed hardware architecture for real-time disparity map computation. Trans Comput 62:690–704

22. Yahav G, Iddan G, Mandelboum D (2007) 3d imaging camera for gaming application. In: 2007 digest of technical papers international conference on consumer electronics, ICCE 2007. IEEE, New York, pp 1–2

23. Yang S, Huang G, Zhao Z, Wang N (2011) Extraction of topographic map elements with SAR stereoscopic measurement. In: 2011 International symposium on image and data fusion (ISIDF). IEEE, New York, pp 1–4

Chapter 9
Real-Time Image Registration via Optical Flow Calculation

In previous chapters, registration parameters of the Panoptic camera are assumed to be known before starting the omnidirectional vision reconstruction. Camera calibration and registration were conducted offline using tools such as the *Camera Calibration Toolbox for Matlab* [9] or Autopano Software [1]. Several shots that were taken from each camera were fed to the toolbox to obtain the intrinsic and extrinsic parameters. Then, omnidirectional vision reconstruction is conducted using the calibration parameters. In this chapter, we will focus on designing an FPGA-based real-time hardware for image registration. We introduce a novel image registration algorithm based on the optical flow calculation. The presented hardware is using a hierarchical block matching-based optical flow algorithm designed for real-time hardware implementation. The algorithm estimates the initial optical flow with block matching and refines the vectors with local smoothness constraints in each level. We evaluate the proposed algorithm with datasets and provide the results compared to ground truth optical flow. Furthermore, we present a reconfigurable hardware architecture of the proposed algorithm for calculating optical flow in real-time. The presented system can process 640 × 480 resolution frames at 39 frames per second (fps). We show that implemented system is one of the best hardware implementations for real-time optical flow calculations.

9.1 Introduction

Image registration is the process of transforming different sets of data into one coordinate system. Datasets might be multiple photographs of the same scene taken at different moments, taken by different type of sensors or even from different point of views. First, one of the images is taken as the reference frame and geometric transformations are applied to align all images with the reference frame. It is frequently used as the first of the other image processing algorithms. Among many

© Springer International Publishing AG 2017
V. Popovic et al., *Design and Implementation of Real-Time Multi-Sensor Vision Systems*, DOI 10.1007/978-3-319-59057-8_9

applications, we can state computer vision, remote sensing, and MRI as the most common ones. It is also used in astrophotography to align images taken of space. There are several surveys on image registration and its applications and some of these surveys can be found in [10, 19, 31, 41].

Optical flow is the apparent motion of brightness patterns observed when a camera is moving relative to the objects being imaged, as stated by Horn [21]. It aims to create a motion field from image sequences assuming that intensity levels are constant among different observations. Different approaches include gradient constraints, phase conservation, block matching, and energy models [4]. It is a well-established topic and has been benchmarked for long time [2, 4, 32]. In optical flow, the goal is to obtain the vector field which describes how the scene is changing over time. It is often needed in a variety of applications, such as robotic navigation [33, 40], moving object detection [13], facial expression recognition [26, 39], orientation estimation [8], segmentation and tracking [34], or driving assistance systems [18].

Among different approaches, the gradient-based methods have less computational complexity and are frequently used in many applications. We can state the Lucas and Kanade method [30] and the Horn and Schunck method [22] as the ones that are computationally inexpensive and frequently used. There are many real-time hardware implementations presented in the literature based on these two methods. In this work, we developed a new algorithm based on hierarchical block matching and local energy minimization. Block matching-based motion estimation is a very mature field and numerous valuable works have been presented throughout the years. These algorithms are widely used for video compression, motion estimation, and compensation. Although both optical flow and traditional block-based motion estimation utilize the constant intensity assumption, the results of motion estimation algorithms do not necessarily correspond to optical flow results. Block matching-based motion estimation techniques try to find out the block most similar to the reference block. The result of this search does not necessarily correspond to the actual motion of the object in the image. In order to overcome the limitations, we will introduce a local energy minimization framework based on the assumption that blocks belong to the same objects should undergo the same motion.

In this chapter, we present an optical flow algorithm based on hierarchical motion estimation and energy minimization suitable for real-time hardware implementation. Furthermore, we will present the design of the reconfigurable hardware implementation of the proposed optical flow algorithm. This algorithm will be used for the image registration step of the real-time multiple frame super-resolution. This chapter is organized as follows: different approaches proposed for the image registration are discussed in Sect. 9.2. Then, the detailed explanation of the optical flow algorithms and current hardware implementations are presented in Sect. 9.3. Development of hierarchical block matching-based optical flow algorithm suitable for real-time hardware implementation is presented in Sect. 9.4.

Hardware implementation of the proposed algorithm and the implementation results are presented in Sect. 9.5. Conclusions and future work are presented in Sect. 9.6.

9.2 Image Registration

Registration is a fundamental task in image processing which is used to match two or more pictures taken at different times, from different sensors, or from different viewpoints. Registration of the images taken by different sensors (such as electro-optical and infrared sensors) is discussed in [23]. For an example of the registration images taken from different viewpoints, we can easily state the *Bullet Time* shots in the movie *Matrix*. Shots taken simultaneously from the different cameras were allowing depth-aware point of view changes. In this work, we focus on images taken by one sensor in different times, or several similar sensors at the same time.

Image registration techniques can be divided into two main approaches, feature-based and intensity-based approaches. We will explain those concepts before explaining the developed optical flow-based algorithm.

9.2.1 Feature-Based Methods

Feature-based algorithms generally consist of the following basic steps:

- **Feature Detection** An image feature is any portion of an image that can potentially be identified and located in both images. Features can be points, lines, or corners. Detecting features can be manual or automatic depending on whether feature detection and matching is human-assisted or performed using an automatic algorithm. SIFT [29] and SURF [6] are famous techniques aimed at recognizing and describing local features automatically.
- **Feature Matching** Once the features are detected, features among the different observations are matched with each other. Various feature descriptors and similarity measures along with spatial relationships among the features are used for that purpose.
- **Transform Model Estimation** Mapping function is constructed after establishing the feature correspondence. It should transform the current image to overlay it on the reference image. For example, the global translation model is one of the most frequently used models. It consists of rotation, translation, and scaling only. This model is often called shape-preserving mapping because it preserves angles and curvatures [41]. However, it assumes that all pixels have the same movement, therefore it cannot handle the local movements, i.e., where several independent movements exist. For many image registration problems, the geometric correspondence between features of the two images is too complex to be characterized by a single transformation function that applies everywhere. For such problems, a transformation function with locally varying parameters may be used. These functions are called local transformations. Contrary to assuming that every pixel in the current frame is globally shifted version of the reference frame, it allows each pixel to move in individual directions. It hassuperior

quality compared to global translation [41]. Furthermore, there are more complex transformation models such as affine and projective transformations, which takes many more parameters into account.

- **Geometric Transformation** Once the transformation model is defined, the current image is transformed to the same coordinate system with the reference image using the transformation metrics. Necessary computations are conducted and non-integer coordinates are computed by the appropriate interpolation technique.

The feature-based techniques can often lead to very good results with high computational complexity. Many methods used to detect corners demand the computation of first and second derivatives of the images. Furthermore, the feature-matching step uses iterative statistical considerations such as RANSAC, which is a method to estimate parameters of a mathematical model from a set of observed data which contains outliers. Due to computational complexity and irregularity of the feature-based algorithms, they are not suitable for real-time hardware implementation.

9.2.2   Intensity-Based Methods

Another method for image registration is finding matching features among the frames using the intensity similarities. This can be achieved by calculating the true motion vectors between the reference and test frame. True motions of each pixel can be utilized for image registration. The majority of motion estimation algorithms in the literature use either block matching or optical flow to determine a dense motion field. Optical flow-based algorithms provide superior motion vector quality over block-based algorithms [2], at the expense of high computational complexity and long run times. In the interest of real-time applications, block matching algorithms provide a flexible trade-off between complexity and MV quality [11]. In this chapter, we introduce a hierarchical block matching-based optical flow algorithm. The algorithm is capable of finding true motion for each pixel in the system, with sub-pixel accuracy. This sub-pixel motion estimation will construct the basis image registration of the multiple frame super-resolution algorithm.

9.3   Optical Flow Estimation

In this section, we will briefly present the state-of-the-art and current benchmarks for optical flow estimation. Furthermore we will discuss the previous hardware implementations before presenting the developed algorithm.

9.3.1 Benchmarking

Since the 1980s, numerous valuable works have been presented for optical flow estimation. Several benchmarks are introduced in order to calculate the quality of these methods. Barron et al. [4] introduced one of the early examples of these benchmarks. These benchmarks have led to rapid and measurable progress for optical flow estimation. Recently Baker et al. [2] presented a new and more challenging dataset with actual ground truth results to push the limits of the current technology. The authors claim that the dataset should ideally consist of complex real scenes with all the artifacts of real sensors (noise, motion blur, etc.). In their website, more than 100 works are competing in order to obtain the best optical flow estimation.

Most of the works available in [2] provide impressive results based on the error metrics. However, these methods are not suitable for hardware implementation. The presented methods are too complex and require heavy computation which is not suitable for real-time hardware implementation. Therefore, most of the hardware implementations in the literature are based on Lucas and Kanade method [30] which was developed in the early 1980s. It is a cost efficient gradient-based algorithm. Thanks to its low complexity, it is frequently used for real-time hardware implementations of optical flow systems.

9.3.2 Hardware Implementations

Most of the previous hardware implementations mainly focus on the Lucas–Kanade (L&K) method [30]. It is a gradient-based local method, based on the assumption that corresponding pixels of an object in different frames have the same degree of brightness. It is one of the most accurate and computationally efficient methods and widely used for optical flow calculation. Due to its nature, it is relatively easy to implement on the hardware level. However, we should state that only small displacements should be present in the scene, so that the first order Taylor approximation is valid.

An early example of a real-time optical flow algorithm is presented in [12]. The authors have implemented the algorithm on a pipelined image processor and performed optical flow calculation of a spatial resolution of 252×316 pixel in 47.8 ms. Díaz et al. [15] presented an example of L&K method hardware implementation on FPGAs. It is capable of processing 320×240 pixel with 30 fps. Ishii et al. [24] presented another example of a high frame rate L&K method implementation on FPGA. The authors developed an optical flow system capable of creating dense motion field with pseudo-variable frame rate. The system is capable of performing 1000 fps with 1024×1024 pixel images. An improved version of L&K algorithm is presented in [16]. The presented work is capable of processing 800×600 pixels at 170 fps.

Seong et al. [37] presented a reduced frame memory access implementation of L&K algorithm by storing the Gaussian filtered operation instead of the original image. The loss in overall system is negligible and the system can process full HD images at 30 fps. Barranco et al. [3] presented a multi-scale implementation of L&K algorithm, capable of processing 32 fps of 640×480 frame size. A different approach for optical flow calculation is presented by Wei et al. [38]. The authors implemented a 3D tensor-based algorithm on an FPGA. The system is capable of processing 640 × 480 images at 64 fps. Another hardware implementation using the algorithm developed by Horn and Schunck is implemented in [20]. The system can calculate the optical flow vector for 256 × 256 pixels in less than 4 ms.

As stated previously, the examples of optical flow hardware systems are mostly based on Lucas and Kanade algorithm. These works try to improve the accuracy of the hardware-based implementation of the method. Furthermore, they provide results using only the Yosemite sequence, not the novel and more complex datasets. In this work, we will present a block matching-based algorithm for optical flow implementation. The presented algorithm provides better results with novel datasets compared to other algorithms and can be implemented in real-time on the hardware level. We will compare the results with actual ground truth results provided by Baker et al. [2].

9.4   Algorithm Development

In this section, we will present a novel algorithm which utilizes the hierarchical sum of absolute difference (SAD)-based block matching and regularization metrics to calculate the optical flow. Hierarchical block matching calculates the initial estimates of the optical flow in the scene. Once the initial estimates are obtained, they are compared according to smoothness metrics. Smoothness metrics aim to state out the vectors that belong to the same object and should undergo the same motion.

9.4.1   Block Matching Algorithms

Block matching is a way of matching macroblocks in a sequence of digital video frames for the purposes of motion estimation. It assumes that the brightness among the two frames is held constant and the corresponding pixels have the same luminance. To make use of this case, frames are divided into previously determined fixed or variable size macroblocks. Each macroblock is compared with other macroblocks in reference frames. The best matching block among the reference blocks is chosen as the candidate corresponding block.

An example of a reference block and a search range is shown in Fig. 9.1. The macroblock being searched is highlighted in Fig. 9.1a and the search range is

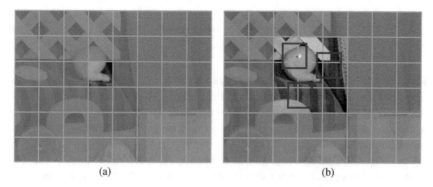

(a)   (b)

Fig. 9.1 The reference (**a**) and the test frames (**b**) for block matching algorithm. The macroblock being search is highlighted in reference frame and search range is highlighted in test frame. Example candidate macroblocks are shown in *red*

highlighted in Fig. 9.1b. Several example macroblocks are also shown in Fig. 9.1b. The goal is to find the best matching macroblock in the search frame using the similarity metrics.

Success of block matching depends on many independent and dependent variables such as block size, search range, initial starting point of the search, and even the direction of the search. For example, using large macroblock sizes leads to incorrect estimation. Individual objects in a macroblock may move independently. A motion vector for large macroblocks cannot represent the motion of each pixel in the block correctly. Furthermore, while using small macroblocks, sufficient texture might not be available. All of the macroblocks might have the same SAD value. The insufficient texture also leads to incorrect estimates. A suitable block size must be chosen according to the needs of the algorithm.

The chosen search range is also an important factor. Large displacements cannot be represented in a small search range, since the SAD-based search will fall into a local minima trap. The local minima trap is the best SAD among the search range but not the correct motion vector. Also, using a large search range might not be feasible in terms of computation time. Each additional range will add significant computational load to the block matching hardware.

Another important aspect that plays an important role in block matching is the search method. Previously, several different search methods have been proposed in the literature. These different methods enable time vs. accuracy trade-offs. Different search methods and their trade-offs are out of the scope of this work. However, to show that solely block matching-based motion estimation is not suitable for optical flow, we will compare optical flow results of different search methods.

Several metrics are proposed to decide the best matching block in the reference frame [14]. Thanks to their nature of robustness and low complexity, correlation-based methods are utilized frequently. The SAD metric which sums the absolute difference of each pixel value can be explained as follows:

$$\text{SAD} = \sum_{i=1}^{N} \sum_{j=1}^{M} |A(i,j) - B(i,j)| \tag{9.1}$$

where $N \times M$ is the chosen block size, usually $N = M$, A and B are the reference macroblock and the search macroblock, respectively. After comparing the reference block with all candidate macroblocks, the one with the lowest SAD result is chosen as the candidate. A vector that relates two macroblocks to each other is defined as the motion vector.

For a quick comparison, we have conducted exhaustive search and two-dimensional logarithmic search-based block matching. Exhaustive search compares all possible candidate macroblocks in the search range. Logarithmic search reduces the computational area and therefore the computational burden [25]. However, it provides lower quality results in comparison to exhaustive search. In the first test, the *RubberWhale* dataset from [2] is used and we compare the obtained results with provided ground truth vectors. Another important metric for motion estimation algorithms is the peak signal to noise ratio (PSNR). It is an approximation to human perception of reconstruction quality [35]. PSNR is frequently used to measure the quality of reconstruction of the lossy compression methods. A motion compensated frame is created with the obtained motion vectors and PSNR values for reference and compensated frames are calculated.

In order to calculate the PSNR, we first need to calculate the mean squared error (MSE). It is the mean of the squared difference between a reconstructed frame and the original frame that wanted to be reconstructed.

$$\text{MSE} = \frac{1}{NM} \sum_{i=1}^{N} \sum_{j=1}^{M} (I(i,j) - K(i,j))^2 \tag{9.2}$$

While PSNR is the ratio of the maximum possible power of a signal to the mean squared error.

$$\text{PSNR} = 20 \log_{10} \left(\frac{\max(I)}{\text{MSE}} \right) \tag{9.3}$$

And for comparing the calculated motion vectors with the ground truth motion vectors, we use the endpoint error (EE) which can be defined as:

$$\text{EE} = \frac{1}{NM} \sum_{i=1}^{N} \sum_{j=1}^{M} \sqrt{\left(u_{i,j} - u_{GT_{i,j}}\right)^2 + \left(v_{i,j} - v_{GT_{i,j}}\right)^2} \tag{9.4}$$

where $u_{GT_{i,j}}$ and $v_{GT_{i,j}}$ are the ground truth x and y dimensions of the motion vectors provided by Baker et al. [2] and $u_{i,j}$ and $v_{i,j}$ are the x and y dimensions found from the conducted search.

Another evaluation method for calculating motion vector accuracy is *angular error*. It is the angular difference between the ground truth vector $\overrightarrow{V_{GT}} = (v_{GT_{i,j}}, u_{GT_{i,j}})$ and calculated motion vector $\overrightarrow{V} = (v_{i,j}, u_{i,j})$. The equation to calculate the angular error is given as:

$$AE = \arccos\left(\frac{1 + u_{GT_{i,j}} \cdot u_{r_{i,j}} + v_{GT_{i,j}} \cdot v_{r_{i,j}}}{\sqrt{1 + u_{r_{i,j}}^2 + v_{r_{i,j}}^2}\sqrt{1 + u_{GT_{i,j}}^2 + v_{GT_{i,j}}^2}}\right) \tag{9.5}$$

Averaging the angular error for the entire frame gives us the *average angular error* (AAE), which is another metric used for evaluation of the computed optical flow vectors.

In Table 9.1, simulation results with different block sizes and different search ranges are provided for block sizes 4×4, 8×8, and 16×16 and for search ranges ± 4, ± 8, and ± 16.

PSNR results from Table 9.1 are as expected, where the results get better with reduced block size and increased search range. On the contrary, endpoint error decreases with increased search range and varies with block size.

Endpoint error results show that using block matching algorithms for true motion estimation is not suitable. Furthermore, block size and search range have significant effects on the endpoint results. In order to minimize the drawbacks of the search range and block size, hierarchical block matching algorithm can be utilized [7].

9.4.2 Hierarchical Block Matching Algorithm

As shown in Table 9.1, exhaustive search method compares the current macroblock with all candidate blocks in the search range. It provides the best results in terms of low SAD and good PSNR compared to other search methods. However, having a

Table 9.1 PSNR (in dB) and EE (in pixels) results for exhaustive search and 2D logarithmic search with different block sizes and search ranges

		±4		±8		±16	
		PSNR	EE	PSNR	EE	PSNR	EE
4	Exh	38.92	0.4638	39.09	0.6103	39.17	0.8598
	Log	38.77	0.4511	38.85	0.5146	38.89	0.5833
8	Exh	37.71	0.3877	37.74	0.4352	37.83	0.51
	Log	37.61	0.391	37.63	0.4039	37.68	0.4584
16	Exh	36.39	0.4174	36.39	0.4174	36.42	0.445
	Log	36.31	0.4222	36.31	0.4222	36.31	0.4222

good PSNR does not always coincide with true motion of the macroblocks. The best PSNR result is obtained with *BlockSize* = 4 and *SearchRange* = ±16. The same parameters also result in the worst endpoint error. SAD-based motion estimation relies on finding the macroblock which is the most similar to the reference block. In reality, similarity in terms of intensity does not always lead to true motion estimation.

As discussed previously, the ability to choose the correct motion vector depends on the chosen block size, search range, initial point of the search, and apparent large motions in the scene. Using large block size creates false matches, where there are several objects moving independently. On the other hand, using small block size causes false local minimum that give good results in intensity comparison-based metrics. However, the resulting vector does not correspond to true motion. For choosing correct block size, search range and search method play significant role in the quality of optical flow. The goal is to obtain good results with spending less computation time while reducing the effects of different options. Hierarchical block matching (HBM) tries to overcome these problems.

The idea is to create an image pyramid and use the results of each level of the pyramid as an initial starting point for the next level [7]. It conducts the motion estimation using different block sizes and search range in each level. In this work, we use three levels of image pyramid.

A three level example of an image pyramid can be seen in Fig. 9.2. The bottom level of the pyramid is obtained by sub-sampling the original image while the top level of the pyramid is obtained by increasing the resolution of the image by interpolation. The middle level is the original image frame.

In the bottom level, large block sizes are used in order to determine a rough estimate of the global motion as an initial estimate for the next level. Once the motion vectors are obtained from the third level, these motion vectors are utilized as a starting point for the second level of the pyramid. The block size is reduced and the search range is updated accordingly. The algorithm continues to perform block matching until the final level of the pyramid is finished. The final level is the interpolated version of the original image for the sub-pixel accuracy.

Hierarchical block matching allows us to utilize different search range and macroblock sizes and tries to overcome the problems posed by each individual choice. However, as discussed earlier, SAD-based search is not sufficient enough for optical flow calculation. For example, we conduct the simple hierarchical block matching algorithm, with block sizes 32, 16, and 8, respectively, in each level. The results of 4×4 macroblocks are assigned to each individual pixel. From the simulations, we obtained the endpoint error as 0.43 and angular error as 11.09°. Although using hierarchical architecture regularizes the motion flow, using solely hierarchical block matching is not sufficient enough. Mainly, the quality of matches at the next level of the hierarchy will strongly depend on the quality of matches at the current level. Therefore, it is necessary to ensure that only valid motion vectors are passed to the next level of the hierarchy. This creates the necessity of a refinement process for the motion vectors at each level before passing them to the next level. In the next section, we will introduce a method for refining the search results which aims to minimize the irregularity between the motion vectors and to decrease the endpoint error.

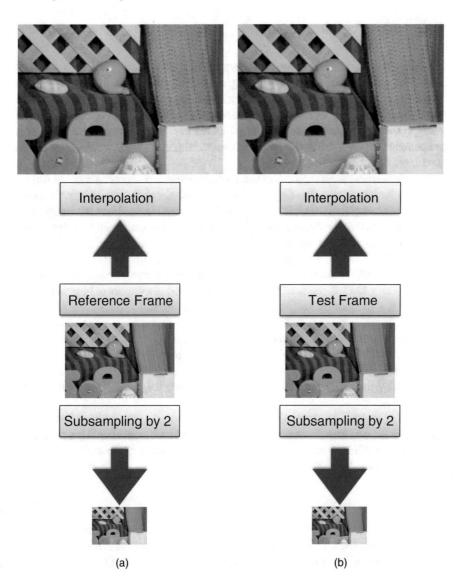

Fig. 9.2 Hierarchical block matching example. (**a**) Reference and (**b**) test frames are obtained from [2]

9.4.3   Flow Regularization

As stated before, results of each step are utilized as a starting point of the next step. The quality of the results is highly dependent on the quality of the motion vectors calculated in the previous step. Incorrect estimates in initial steps are passed throughout the levels and the error accumulates as the motion vectors are passed through the levels. Before passing the estimated motion vectors to the next level, a refinement process can help to detect incorrect estimates and reduce the error on the next level of the hierarchy.

The fundamental assumption of the motion estimation algorithms is the brightness consistency among the displacements. It assumes that pixels retain the same luminance among their displacement path. Regularization methods introduce additional constraints to the motion estimation framework to solve the ill posed problem. The introduced methods are based on another assumption that blocks which belong to same objects should undergo the same displacement, thus have the same motion vector. These additional constraints are called the smoothness constraints [5]. These constraints try to limit the solution based on previous assumptions. In general, it is assumed that the motion field is locally constant or linear.

The refinement process should state out the incorrect motion estimation vectors that belong to the same motion field and propose a new motion vector instead. It should penalize the deviation of the original motion vector $\overrightarrow{mv}_{\text{ori}}$ from its neighbor vectors $\overrightarrow{mv}_{1...N}$. However, we should keep in mind that the new motion vector proposed by the refinement process might not have the same SAD value as the previous vector.

In order to find the optimal motion vector, we merge smoothness-based constraints to the previously discussed gradient-based block matching. This process is called energy minimization. It tries to minimize the energy of the motion field with previously defined energy function. The defined energy function is as follows:

$$E_i = \text{SAD}(I_0, I_1, \mathbf{v}_i) + \lambda S(\mathbf{v}_i) \tag{9.6}$$

where $\text{SAD}(I_0, I_1, \mathbf{v}_i)$ is the SAD value coming from the block matching algorithm, $S(\mathbf{v}_i)$ is the smoothness factor aiming to penalize the deviations in the motion field, and λ is the weight to regularize the smoothness vector over SAD results.

The idea of the smoothness term is to compare the calculated motion vector with its neighbors and understand if those vectors actually correspond to same motion block. First, we should determine how many motion vectors should be selected for comparison. First order and second order neighbors can be seen in Fig. 9.3. We found out that using second order neighbors are giving better results than the first order neighbors, without adding significant computation load on the hardware architecture. Therefore, we have selected to use the second eight neighbors during the smoothness calculation.

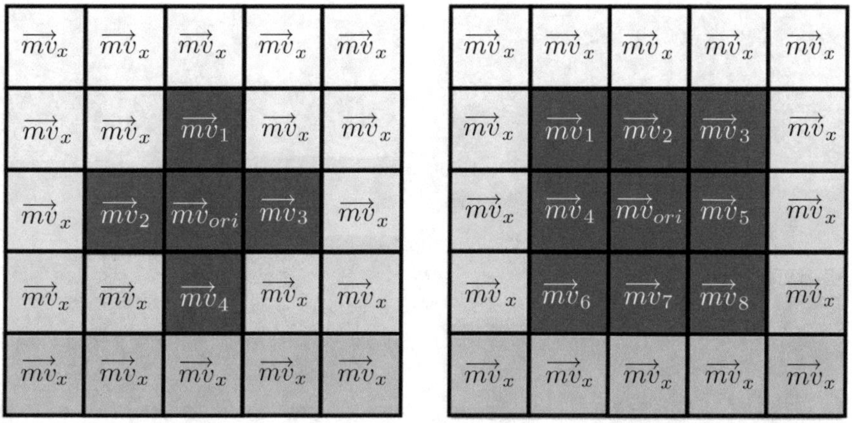

Fig. 9.3 (a) First and (b) second order neighbors for smoothness constraint calculations

Second, we have to decide a penalty function to penalize the deviations among the motion block. Previously, several penalty functions including l_1 and l_2 norms have been introduced as low-complexity smoothness constraints [5]. We have decided to use l_2 norm of the second order neighborhood. Although implementation of the l_1 norm has a low complexity over the l_2 norm, the l_2 norm provided better results and we have decided to continue on the l_2 norm implementation.

In [36], the λ value that weights the smoothness over the SAD value is called Lagrange multiplier and the authors stated that they used λ as twice the block size. However, throughout the simulations, we found out using $\lambda = BlockSize$ gives better results in terms of endpoint error, therefore we decided to fix λ equal to *BlockSize*.

While computing the energy, we need to keep in mind that the energy of the vector $\overrightarrow{mv}_i$ might only be less than the energy value of the vector $\overrightarrow{mv}_{ori}$, if the smoothness value for the vector $\overrightarrow{mv}_i$ is less than the smoothness value of the $\overrightarrow{mv}_{ori}$. Since the block matching algorithm has already computed the SAD value of the each block in the search range, we know that SAD value of the block represented with $\overrightarrow{mv}_i$ is greater than or equal to SAD value of the block represented with $\overrightarrow{mv}_{ori}$. Once we compute all the smoothness values, we only need to calculate the energy of $\overrightarrow{mv}_i$ E_i if the smoothness value is less than the smoothness value of $\overrightarrow{mv}_{ori}$.

Example results of flow regularization can be seen in Fig. 9.4. Motion vectors for the second hierarchical level of the *RubberWhale* dataset can be seen in Fig. 9.4a. After scanning the motion vectors and energy calculation, updated motion vectors are depicted in Fig. 9.4b. Many irregular motion vectors have been updated with regular lower energy motion vectors.

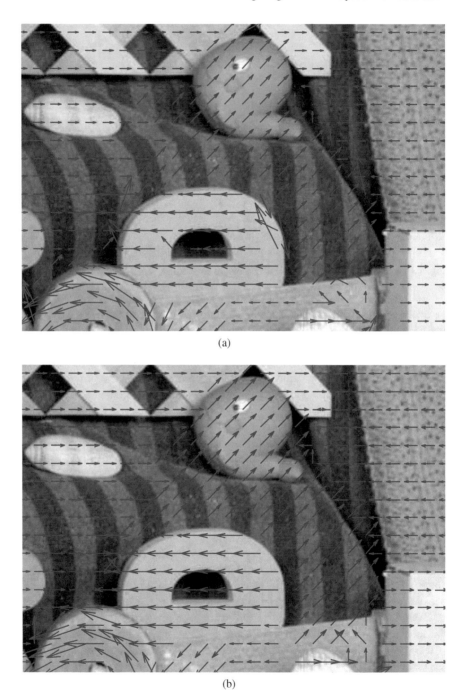

(a)

(b)

Fig. 9.4 Motion vectors (**a**) before and (**b**) after flow regularization. Examples shown in figures are taken from the second hierarchical level

---

Algorithm 4 Proposed algorithm for optical flow

1: Create Image Pyramid
2: **for** *Levels* 1 **to** 3 **do**
3:     Conduct Block Matching
4:     **for** each $\overrightarrow{mv}$ **do**
5:         **for** *Neighbours* 1 **to** 8 **do**
6:             **if** $\overrightarrow{mv}_{ori} \neq \overrightarrow{mv}_i$ **then**
7:                 Calculate E_i and E_{ori}
8:                 **if** $E_i < E_{ori}$ **then**
9:                     Update $\overrightarrow{mv}_{ori}$ as $\overrightarrow{mv}_i$
10:                **end if**
11:            **end if**
12:        **end for**
13:    **end for**
14:    Pass motion vectors to next level
15: **end for**

9.4.4 Proposed Algorithm

In this section we will present the final algorithm for optical flow estimation (see Algorithm 4). After the image pyramid is created, SAD-based block matching is first conducted for the lowest resolution image. Once the block matching results are obtained, the motion vector field is refined before passing to the next level.

For each motion vector, the second order neighbors as seen in Fig. 9.3b are compared. Only the vector values which are different from the original vector value $\overrightarrow{mv}_{\mathrm{ori}}$ are taken into account. If all motion vectors are equal, no further action is needed. If not, the energy of the original motion vector E_{ori} is first calculated as explained in Eq. (9.6). Afterwards the $\overrightarrow{mv}_{\mathrm{ori}}$ is replaced with $\overrightarrow{mv}_i$ and energy E_i of the motion vector is calculated again. If the energy of the substituted vector E_i is better than the energy of the original motion vector E_{ori}, $\overrightarrow{mv}_{\mathrm{ori}}$ is substituted with $\overrightarrow{mv}_i$ and the motion field is updated. Updated motion estimation results are passed on to the next level of block matching algorithm and used as initial starting points. The algorithm is conducted until the energy minimization of the final level is finished.

The quality of the results also depends on the block sizes and search ranges in each level. In order to handle both large and small displacements in the datasets, we found out that using block sizes as 32, 16, and 8, respectively, for each level provides reliable results for image size 640×480. Block sizes less than 8×8 decrease overall quality of matching, which is also consistent with Table 9.1. Since the block size is reduced into half while passing to next level, each motion vector in the current level of hierarchy will be used as the starting point of the 16 macroblocks in the next level.

Finally, search ranges among the hierarchy levels can vary. Through the simulations, we concluded that ± 16 for lower resolution in the pyramid, ± 8 for original image, and ± 4 for interpolated image give good endpoint error results.

Table 9.2 Endpoint error (in pixels) and angular error (in degrees) comparison for Lucas and Kanade, Horn and Schunck, and this work

	L&K		H&S		This work	
	EE	AE	EE	AE	EE	AE
RubberWhale	0.37	10.87°	0.55	15.00°	0.28	8.59°
Dimetrodon	1.54	39.36°	1.00	19.01°	0.44	8.23°
Venus	3.47	54.74°	3.63	43.89°	0.47	6.41°
Hydrangea	3.09	50.25°	3.18	40.84°	1.98	14.80°
Grove2	2.79	59.30°	2.65	40.48°	0.42	5.80°
Grove3	3.47	50.37°	4.41	40.29°	0.99	10.90°

9.4.5   Results

Endpoint error and angular error results for the proposed algorithm, as well as Lucas and Kanade and Horn and Schunck methods are presented in Table 9.2. The results are compared with ground truth optical flow provided by Baker et al. [2]. None of these methods are calculating the occluded pixels, therefore the occluded pixels provided in ground truth vectors are taken out while calculating the endpoint and angular error. While calculating the motion vectors, parameters for the proposed algorithm are kept fixed and motion vector representing each 4×4 block is assigned to 16 corresponding pixel in the macro block. For other methods, we used the toolbox provided in [17] and made a parametric sweep to find the best results that can be obtained from the algorithms.

An example portion of the results for comparing the *RubberWhale* sequence is shown in Fig. 9.5. From the images, we can see that the proposed method leads to a regular optical flow in comparison to other methods and provides better results. The obtained results are not good with the scenes that have complex movements such as *Hydrangea* dataset. In contrary, the results are better with the scenes that have close to real-life scene and movements, such as *Dimetrodon* or *RubberWhale*.

From Table 9.2 and Fig. 9.5, we conclude that the proposed algorithm provides better angular error and endpoint error results compared to the other methods that are frequently used for real-time hardware implementations in all the datasets. In the next section, we will present the real-time hardware implementation of the proposed algorithm.

9.5   Hardware Implementation of the Proposed Algorithm

In this section, we will explain the implementation of the proposed optical flow algorithm in Sect. 9.4. For the block matching unit, we have implemented a widely used 2d systolic array architecture and we have designed an additional circuit for smoothness calculation. With addition of a few reconfigurable units, the

(a)   (b)

(c)   (d)

Fig. 9.5 Zoomed in portion of the optical flow estimation of the *RubberWhale* sequence. Motion vectors are subsampled for a better visualization of the results. (**a**) Ground truth. (**b**) Proposed algorithm. (**c**) Lucas & Kanade. (**d**) Horn & Schunck

same structure for block matching unit can be used in all three hierarchy levels. Furthermore reconfigurability allows us to calculate SAD results in parallel for the second and the third level of the pyramid.

9.5.1 Block Matching Unit

For the block matching algorithm, a 2D systolic array-based SAD calculator is implemented [27]. The detailed timing diagram and explanation can be found in [28]. An 8 × 8 version of the implemented 2D systolic array can be seen in Fig. 9.6. **AD** blocks consist of absolute difference calculators and adders. Firstly, the reference frame data is loaded into the blocks. The **AD** block calculates the absolute difference of the input search frame data. Afterwards it sums up the values coming from the previous level and passes the result to the next level in the systolic array. **A** blocks sum up all the column addition results. The **M** block in the end holds the addition results for each macroblock. A control hardware is designed to obtain

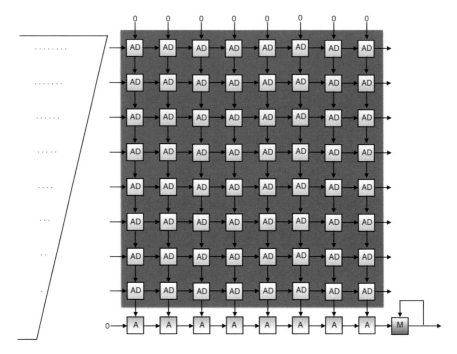

Fig. 9.6 An 8 × 8 2D systolic array architecture for block matching algorithm

valid motion estimation results from the invalid results. The control logic also holds the minimum SAD value and its representing motion vector.

The black dots represent the pipelined values of the input search frame. Reference block data remains fixed in the processing elements (PEs). Search area is provided in a delayed manner. Data providing diagram can also be seen in Fig. 9.6. Each line is one clock cycle delayed from the previous line.

The overall delay of the block matching algorithm depends on the *BlockSize*, *BS* and the *SearchArea*, *SA*. The number of clock cycles needed for one block can be calculated as follows:

$$ClockCycle = BS(SA - BS + 1) \tag{9.7}$$

For our case, with $BS = 32$, $SR = 16$, and $SA = 64$, each block for the lowest level of the pyramid will take 1056 cycles. It will take 272 in level 2 and 72 in level 1. For the final level of the pyramid, in order to obtain the sub-pixel accuracy, a linear interpolation unit has been designed and results of the block are fed into data feeder unit.

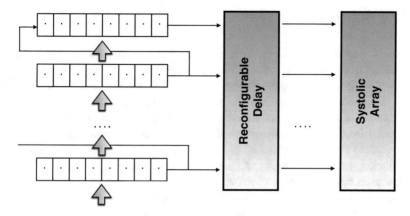

Fig. 9.7 Data feeder block designed for providing data for 2D systolic array

9.5.2 Data Feeder Unit

Another important aspect is to load the reference frame data to systolic array block. One way is to have *BlockSize* × *SearchRange* sized shift registers and load data in serial. However, loading each search frame will take $SR \times BS$ cycles. This is a huge delay, especially when compared to the 2D systolic array architecture delay discussed in Sect. 9.5.1.

For the data providing operation, a shift register-based loading block is designed. The designed block is depicted in Fig. 9.7. The block works in a way that it accepts parallel load of the data. Once the data is loaded, the system shifts the data registers one by one to the next level. In this way, a much faster data loading scheme is achieved. The first input of the reference frame data to systolic array can be provided in the 33rd clock cycle.

Images are stored in 8 bit mono-color format in the external memory of the Xilinx VC707 FPGA. The memory bandwidth is 512 bits per read operation. Therefore, image data read from the memory is first saved in local FIFOs in 512 bits format, which corresponds to 64 pixels. Once the data filled in the FIFOs, search frame data is loaded into the data feeder block in parallel. Upcoming frame data are read from the memory while the current data feeder block is shifting data in the systolic array.

For example, in the 3rd level of the pyramid, data read from the FIFO are first loaded in parallel for 32 cycles. Then in the next 64 cycles, data are shifted through the shift registers. Once a line is finished, the next line is loaded in parallel to the data feeder block, while all other lines are shifted up. This operation continues until the search is finished. For the 3rd level of the pyramid, it only takes 32 cycles to load the data to the block matching block.

Another block between the data feeder block and systolic array block is designed such that it can provide reconfigurable delay time for each line. We will discuss the reconfigurability in the upcoming sections. For the moment, it can be assumed that the delay is 0 cycle for the first line and 31 cycles for the last line.

9.5.3   Regularization Block

As stated previously, we have decided to use the l_2 norm instead of the l_1 norm for smoothness constraint calculation. Although the calculation for the l_2 norm is more complex and requires more resources, it provides better motion field. Multiplier units have instantiated to obtain fast multiplication results. For square root operation, a look-up table-based square root operation has been implemented. The precision loss in the operation does not affect the final results which has been verified via *Matlab* simulations. In total, it took seven cycles to compute one smoothness result or 8 second order neighbors indicated in Fig. 9.3b.

A simple motion equality check mechanism is implemented in order to prevent re-computation of the same motion vector smoothness values. The smoothness is only calculated if the motion vector $\overrightarrow{mv}_i$ appears for the first time in smoothness calculation block and it is different than the $\overrightarrow{mv}_{\mathrm{ori}}$. For the *RubberWhale* dataset that we are working on, we calculated that on average 2.8 vectors are different in each smoothness calculation cycle. This means instead of eight calculations, we need three calculations on average.

Recalling the smoothness value equation of Eq. (9.6), the SAD value of the new block also needs to be calculated. For calculating a single block SAD value, a 2D systolic array architecture is not necessarily needed. Therefore a 1D architecture is implemented for calculating the new SAD value for the motion vector $\overrightarrow{mv}_i$.

9.5.4   Hardware Reuse and Reconfigurability

For implementing different hierarchical levels, one can create three different 2D systolic arrays similar to the one depicted in Fig. 9.6. However, we can use several small systolic array architectures and use them depending on the needs of the current level. An example of such system is depicted in Fig. 9.8. Small block depicts the 8×8 systolic array absolute difference calculator array highlighted in Fig. 9.6. By connecting 16 small systolic arrays, we can create one 32×32, four 16×16, and sixteen 8×8 systolic arrays.

Similarly, the data feeder block in Fig. 9.7 is divided into smaller blocks and modified to be reconfigurable such that depending on the reference frame size and search range size, the same unit can be used in each level of hierarchy. The reconfigurable data provider block can be seen in Fig. 9.9. Shift registers can be reconfigured to shift 64, 32, or 16 data for the search frame. Furthermore, data can be provided directly 1st, 9th, 17th, or 25th line so that no clock cycle will be lost in the meantime. The first data for second level of the hierarchy will be provided in 17th cycle and 9th cycle for the final level.

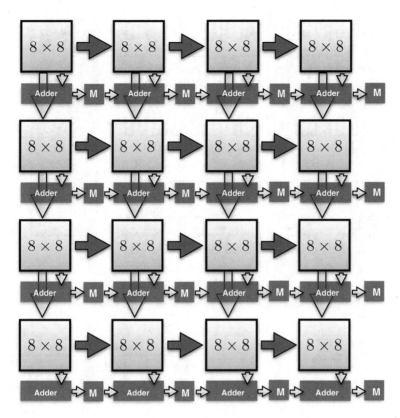

Fig. 9.8 A 32 × 32 systolic array consists of 16 8 × 8 systolic arrays

9.5.5   Timing Analysis

In this section, we will discuss the worst case scenario and provide additional improvements to obtain faster results.

The equation for calculating the number of clock cycles needed for the 2D systolic array is discussed in Sect. 9.5.1 and the results are provided by Eq. (9.7). For each level, it needs 1056, 272, and 72 cycles, respectively. The data feeder block aims to load the data in 32, 16, and 8 cycles. Therefore one block for the hierarchy level one needs 1088 cycles, the second level needs 288 cycles, and 80 cycles is needed for the last level.

The aim is to provide real-time half-pixel accuracy results for VGA(640 × 480) image resolution. The number of blocks and the number of cycles needed for each block as well as the total cycles are provided in Table 9.3. In total, block matching for the single frame can be finished in less than $2M$ cycles.

We have indicated that although there are eight smoothness vectors of the neighbors to be calculated, calculating the smoothness value of all the eight vectors

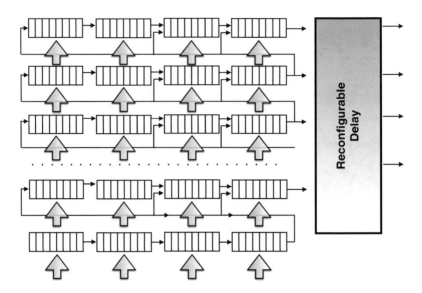

Fig. 9.9 Reconfigurable data providing block, same block can be used for all hierarchical levels of the pyramid

Table 9.3 Number of clock cycles needed for each hierarchy level for block matching part

Level	Blocks	Clock cycle	Total
Level 1	80	1096	87,040
Level 2	1200	288	345,600
Level 3	19,200	80	1,536,000

is not necessary. For this particular reason, smoothness value calculation for the equal vectors is skipped. However, in this analysis, we will assume that all the vectors are different for each neighboring vectors.

In order to calculate the smoothness constraint of each vector, we need seven cycles. After calculating nine values and finding the minimum smoothness value, we can calculate the energy of the new vector. If the new energy value is less than the original energy value, we will update the motion vector with the new motion vector.

For calculating the smoothness value of all vectors, nine motion vectors are read from the block rams. Afterwards $9 \times 7 = 63$ cycles are needed to calculate the smoothness values for each vector. For calculating the SAD value, we need 1024 cycles. In total we need 1096 cycles for the top level of the pyramid. Similarly, we need 328 cycles for the middle level and 136 cycles for the final level. The total number of cycles needed for each block can be seen in Table 9.4. In total, we need less than $3.1M$ cycles for energy minimization.

For the overall system, assuming each step is started right after the previous step, $5.1M$ cycles are needed for each frame to be processed and obtain the optical flow vectors.

Table 9.4 Number of clock cycles needed for each hierarchy level for energy minimization part

Level	Blocks	Clock cycle	Total
Level 1	80	1096	87,680
Level 2	1200	328	393,600
Level 3	19,200	136	2,611,200

9.5.6 Implementation Results

The defined system is implemented using VHDL on a *Xilinx* Virtex 7 XC7VX485T FPGA. The system is designed to work with a 200 MHz clock rate. Each individual block is synthesized, placed, and routed with the Vivado Design Suite and verified that the system can work with the target clock frequency. Since each frame is taking $5.1M$ cycles, the proposed system can process up to 39 frames per second (fps).

The utilized FPGA system has 303,600 Slice LUTs and 607,200 Slice Registers. The implemented hardware utilizes 12% of Slice LUTs and 7% of Slice Registers.

9.6 Conclusion and Future Work

This work proposes a hierarchical block matching-based optical flow calculation algorithm, suitable for real-time hardware implementation. The algorithm utilizes two assumptions while calculating the optical flow, (1) constant intensity assumption, that the pixel intensity levels between the observations are kept constant and (2) pixels belonging to the same object should undergo same motion between observations. By using these two assumptions, the algorithm tries to minimize the energy function defined in Eq. (9.6) among the motion vectors. The energy function tries to reduce the local motion variations in between the neighboring motion vectors. Experiments showed that the algorithm can provide better results with current complex datasets in comparison to other works frequently used for hardware implementations.

In the second part, the implemented hardware system of the proposed algorithm is presented. The proposed system uses a 2D reconfigurable systolic array for real-time implementation of the SAD-based block matching algorithm. Using the same systolic array in each hierarchical level reduces the resource utilization rate. Additionally, another block for calculating the energy is designed. Complex calculations such as square root operation are implemented as LUT-based hardware. This results in lower resource utilization without any loss in the performance.

The system is capable of calculating optical flow with half-pixel precision for 4×4 sized blocks. Results for each block are assigned to each pixel as their optical flow result. For more precise calculation of the optical flow, another level of hierarchy can be added such that the system provides quarter-pixel accuracy. However, this will increase the computation time significantly.

Currently, reconfigurable hardware can process two 16×16 SAD searches and four 8×8 SAD searches, thanks to its reconfigurable nature. However, currently only one search can be executed in each level. As a future work, it is planned to use each sub-block in parallel so that SAD-based search can be conducted much faster. Furthermore, the energy minimization block needs only three lines of the motion vector to start the computation. In this work, the SAD block and energy minimization block do not work at the same time. As a future work, it is planned to start the energy minimization when there are 3 lines of motion vectors available in each level. This will reduce the total number of clock cycles needed for the overall system and can boost the performance of the system up to 60 fps.

The optical flow-based image registration hardware presented in this chapter will be used to register the images for the super-resolution algorithm presented in the next chapter. One important feature that can be added to the calculated optical flow motion vectors is adding a confidence factor to each vector. The confidence factor is correlated to the energy calculated during the flow regularization step. Occluded blocks can be pointed out using the confidence factor. Furthermore, low confidence pixel blocks can be taken out from the super-resolution implementation. We will discuss the multiple frame super-resolution algorithm in the next chapter.

References

1. Autopano (2016) http://www.kolor.com. Accessed on 10 Jan 2016
2. Baker S, Scharstein D, Lewis J, Roth S, Black MJ, Szeliski R (2011) A database and evaluation methodology for optical flow. Int J Comput Vis 92(1):1–31
3. Barranco F, Tomasi M, Diaz J, Vanegas M, Ros E (2012) Parallel architecture for hierarchical optical flow estimation based on FPGA. IEEE Trans Very Large Scale Integr VLSI Syst 20(6):1058–1067
4. Barron JL, Fleet DJ, Beauchemin SS (1994) Performance of optical flow techniques. Int J Comput Vis 12(1):43–77
5. Bartels C, De Haan G (2010) Smoothness constraints in recursive search motion estimation for picture rate conversion. IEEE Trans Circuits Syst Video Technol 20(10):1310–1319
6. Bay H, Tuytelaars T, Van Gool L (2006) Surf: speeded up robust features. In: Computer vision–ECCV 2006. Springer, New York, pp 404–417
7. Bierling M (1988) Displacement estimation by hierarchical blockmatching. In: Visual communications and image processing'88: third in a series. International Society for Optics and Photonics, Bellingham, WA, pp 942–953
8. Bigün J, Granlund GH, Wiklund J (1991) Multidimensional orientation estimation with applications to texture analysis and optical flow. IEEE Trans Pattern Anal Mach Intell 13(8):775–790
9. Bouguet J (2011) Camera calibration toolbox for matlab. [Online] Available: http://www.vision.caltech.edu/bouguetj/calib_doc. Accessed on 7 Dec 2011
10. Brown LG (1992) A survey of image registration techniques. ACM Comput Surv 24(4):325–376
11. Callico G, Lopez S, Sosa O, Lopez J, Sarmiento R (2008) Analysis of fast block matching motion estimation algorithms for video super-resolution systems. IEEE Trans Consum Electron 54(3):1430–1438. doi:10.1109/TCE.2008.4637637

12. Correia MV, Campilho AC (2002) Real-time implementation of an optical flow algorithm. In: Proceedings of 16th international conference on pattern recognition, 2002, vol 4. IEEE, New York, pp 247–250
13. Cucchiara R, Grana C, Piccardi M, Prati A (2003) Detecting moving objects, ghosts, and shadows in video streams. IEEE Trans Pattern Anal Mach Intell 25(10):1337–1342
14. De Haan G (2004) Video processing. University Press, Eindhoven
15. Díaz J, Ros E, Pelayo F, Ortigosa EM, Mota S (2006) FPGA-based real-time optical-flow system. IEEE Trans Circuits Syst Video Technol 16(2):274–279
16. Díaz J, Ros E, Agís R, Bernier JL (2008) Superpipelined high-performance optical-flow computation architecture. Comput Vis Image Underst 112(3):262–273
17. Dollár P. Piotr's computer vision matlab toolbox (PMT). http://vision.ucsd.edu/~pdollar/toolbox/doc/index.html
18. Frenz H, Lappe M, Kolesnik M, Bührmann T (2007) Estimation of travel distance from visual motion in virtual environments. ACM Trans Appl Perception 4(1):3
19. Gholipour A, Kehtarnavaz N, Briggs R, Devous M, Gopinath K (2007) Brain functional localization: a survey of image registration techniques. IEEE Trans Med Imaging 26(4):427–451
20. Gultekin GK, Saranli A (2013) An FPGA based high performance optical flow hardware design for computer vision applications. Microprocess Microsyst 37(3):270–286
21. Horn B (1986) Robot vision. MIT, Cambridge
22. Horn BK, Schunck BG (1981) Determining optical flow. In: 1981 Technical symposium east. International Society for Optics and Photonics, Bellingham, WA, pp 319–331
23. Irani M, Anandan P (1998) Robust multi-sensor image alignment. In: Sixth international conference on computer vision, 1998. IEEE, New York, pp 959–966
24. Ishii I, Taniguchi T, Yamamoto K, Takaki T (2012) High-frame-rate optical flow system. IEEE Trans Circuits Syst Video Technol 22(1):105–112
25. Jain JR, Jain AK (1981) Displacement measurement and its application in interframe image coding. IEEE Trans Commun 29(12):1799–1808
26. Kenji M (1991) Recognition of facial expression from optical flow. IEICE Trans Inf Syst 74(10):3474–3483
27. Komarek T, Pirsch P (1989) Array architectures for block matching algorithms. IEEE Trans Circuits Syst 36(10):1301–1308
28. Kuhn PM, Kuhn Peter M (1999) Algorithms, complexity analysis and VLSI architectures for MPEG-4 motion estimation, vol 8. Springer, Berlin
29. Lowe DG (2004) Distinctive image features from scale-invariant keypoints. Int J Comput Vis 60(2):91–110
30. Lucas BD, Kanade T et al (1981) An iterative image registration technique with an application to stereo vision. In: IJCAI, vol 81, pp 674–679
31. Maintz JA, Viergever MA (1998) A survey of medical image registration. Med Image Anal 2(1):1–36
32. McCane B, Novins K, Crannitch D, Galvin B (2001) On benchmarking optical flow. Comput Vis Image Underst 84(1):126–143
33. McCarthy C, Bames N (2004) Performance of optical flow techniques for indoor navigation with a mobile robot. In: Proceedings of 2004 IEEE international conference on robotics and automation, 2004, vol 5. ICRA'04. IEEE, New York, pp 5093–5098
34. Mikić I, Krucinski S, Thomas JD (1998) Segmentation and tracking in echocardiographic sequences: active contours guided by optical flow estimates. IEEE Trans Med Imaging 17(2):274–284
35. Richardson I (2003) H.264 and MPEG-4 video compression: video coding for next-generation multimedia. Wiley, New York. https://books.google.it/books?id=ECVV_G_qsxUC
36. Santoro M, Al-Regib G, Altunbasak Y (2012) Adaptive search-based hierarchical motion estimation using spatial priors. In: VISAPP (2), pp 399–403
37. Seong H, Rhee C, Lee H (2015) A novel hardware architecture of the Lucas-Kanade optical flow for reduced frame memory access. IEEE Trans Circuits Syst Video Technol (99):1–1. doi:10.1109/TCSVT.2015.2437077

38. Wei Z, Lee DJ, Nelson BE (2007) FPGA-based real-time optical flow algorithm design and implementation. J Multimed 2(5):38–45
39. Yacoob Y, Davis LS (1996) Recognizing human facial expressions from long image sequences using optical flow. IEEE Trans Pattern Anal Mach Intell 18(6):636–642
40. Zingg S, Scaramuzza D, Weiss S, Siegwart R (2010) Mav navigation through indoor corridors using optical flow. In: 2010 IEEE international conference on robotics and automation (ICRA). IEEE, New York, pp 3361–3368
41. Zitova B, Flusser J (2003) Image registration methods: a survey. Image Vis Comput 21(11):977–1000

Chapter 10
Computational Imaging Applications

The previous part of the book introduced five multi-camera systems, their design, and FPGA implementation for real-time omnidirectional video construction in both low and high resolutions. This chapter will present several developed applications of multi-camera systems. The first application is the HDR imaging implemented on a platform Panoptic Media Platform, described in Chap. 6, but it is easily portable to any other multi-camera system with large FOV overlap between the cameras. The other applications are based on depth estimation hardware and include hardware-based free-view synthesis, as well as multiple real-time software applications that use FPGA-generated depth map. The implemented applications conceptually prove that the high-quality and high-performance RGB+D outputs of the proposed real-time disparity estimation hardware can be used for enhanced 3D-based video processing applications.

10.1 High Dynamic Range Imaging

10.1.1 Introduction

Dynamic range in the digitally acquired images is defined as the ratio between the brightest and the darkest pixel in the image. Most modern cameras cannot capture sufficiently wide dynamic range to truthfully represent radiance of the natural scenes, which may contain several orders of magnitude from light to dark regions. This results in underexposed or overexposed regions in the taken image and the lack of local contrast. Figure 10.1 shows three shots taken under different exposure settings of a camera. The underexposed and overexposed images show fine details in very bright and very dark areas, respectively. These details cannot be observed in the moderately exposed image.

© Springer International Publishing AG 2017
V. Popovic et al., *Design and Implementation of Real-Time Multi-Sensor Vision Systems*, DOI 10.1007/978-3-319-59057-8_10

(a)   (b)   (c)

Fig. 10.1 A subset of images taken for recovering the camera response curve. The images are taken with (**a**) short, (**b**) medium, and (**c**) long exposure time

HDR imaging technique was introduced to increase dynamic range of the captured images. HDR imaging is used in many applications, such as remote sensing [8], biomedical imaging [21], and photography [5], thanks to the improved visibility and accurate detail representation in both dark and bright areas.

HDR imaging relies on encoding images with higher precision than standard 24-bit RGB. The most common method of obtaining HDR images is called exposure bracketing and it includes taking several low dynamic range (LDR) images, all under different exposures [27]. Debevec and Malik [10] developed an algorithm for creating wide range radiance maps from multiple LDR images. The algorithm included obtaining camera response curve, creation of HDR radiance map, and storage in RGBE format [46]. Other approaches based on a weighted average of differently exposed images were proposed in [32, 34, 39], with differently calculated weights. State-of-the-art algorithms for radiance map construction include camera noise model and optimization of the noise variance as the objective function [14, 17].

An exposure bracketed sequence can also be fused into the HDR image without the radiance map calculation. Exposure fusion method [30] is a pipelined fusion process where LDR images are combined based on saturation and contrast quality metrics. Thanks to direct image fusion, the exposure fusion is not a resource-demanding algorithm as there is no HDR radiance map to be stored, which significantly reduces the memory requirement. A similar principle is used for contrast enhancement using a single LDR image [40]. An alternate option to exposure bracketing is to use an adjustable camera response curve sensor, such as LinLog [29].

Besides capturing the natural scenes, another problem occurs when displaying them. The modern displays are limited to the low dynamic range, which causes inadequate representation of even standard LDR images. In order to avoid such problems, a tone mapping operation is introduced to map the real pixel values to the ones adapted to the display device. The purpose of tone mapping is to compress the full dynamic range in the HDR image, while preserving natural features of the scene.

Tone mapping operators can be divided into two main groups named global and local operators. Global operators are spatially invariant because they apply the same transformation to each pixel in the image. These algorithms usually have low complexity and high computational speed. However, such algorithms have problems preserving the local contrast in the images where the luminance is uniformly occupying the full dynamic range. The first complex global techniques were based on human visual system (HVS) model and subjective experiments [33, 47]. The latest global techniques are based on adaptive mapping. Drago et al. [11] introduced an adaptive logarithmic mapping which applies different mapping curves based on pixel luminosity. The curves vary from $\log_2$ for the darkest pixels, to $\log_{10}$ for the brightest. Similarly, Mantiuk et al. [28] have recently developed a tone mapping algorithm adaptive to the display device.

Opposite to the global operators, local operators are more flexible and adaptable to the image content, which may drastically improve local contrast in regions of interest. Since they differently operate on different regions of the image, they are computationally more expensive and resource-demanding. Reinhard et al. [38] introduced a local adaptation of a global logarithmic mapping. The adaptation was inspired by photographic film development in order to avoid halo artifacts. Fattal et al. [13] proposed an operator in gradient domain which was computationally more efficient than other local operators. Nevertheless, both Reinhard and Fattal operators are very resource-demanding for large images, since they require a Gaussian pyramid decomposition and a Poisson equation solver, respectively. Durand and Dorsey [12] presented a fast bilateral filtering where high contrast areas are preserved in the lower spatial frequencies. However, the main disadvantage of this method is the significantly lower overall brightness.

Obtaining and reproducing the HDR video is a difficult challenge due to various issues. Majority of the techniques use exposure bracketed frames from a single camera, which results in high motion blur among frames. Furthermore, using frames from the same camera inherently lowers the effective frame rate of the system, independently of the tone mapping process. The display frame rate is further influenced by both the HDR imaging technique and the processing system. Majority of the systems are based on CPU or GPU. Even though GPUs are targeted to process large amount of data in parallel, they often fail to meet the tight real-time timing constraints.

This section presents simultaneous real-time HDR video construction and rendering using a multi-camera system [41]. The key idea is to use a multi-camera setup to create a composite frame, where cameras with the overlapping FOV are set to different exposure times. Such system reduces the motion blur, as there is no inter-frame gap time (which can be several hundred milliseconds in the standard HDR cameras). Additionally, the frames are captured at the same moment by all cameras, which reduces the intra-frame motion of the scene objects to the difference interval of cameras' exposure times.

10.1.2 Related Work

Exposure bracketing using a single video camera is the most widely used method for HDR video construction. Kang et al. [23] proposed a method of creating a video from an image sequence captured while rapidly alternating the exposure of each frame. Kalantari et al. [22] apply the identical principle and use the patch-based synthesis to deal with the fast movements in the scene. The HDR construction in both cases is realized in post-processing and does not have the real-time processing capability. Gupta et al. [15] recently proposed a way of creating HDR video using Fibonacci exposure bracketing. In this work they adapted a machine vision camera Miro M310 to quickly change exposures and thus reduce the inter-frame delay. However, this system still requires significantly long time to acquire a sequence of frames with desired exposures. Another approach is to use a complex camera with beam splitters [43]. A similar setup is also used in the work of Kronander et al. [24]. This spatially adaptive HDR reconstruction algorithm fits local polynomial approximation to the raw sensor data. However, the algorithm requires intensive processing to recover and display the HDR video.

Exposure bracketing can also be used in multi-camera or multi-view setups. Ramachandra et al. [37] proposed a method for HDR deblurring using already captured multi-view videos with different exposure times. Portz et al. [36] presented a high-speed HDR video using random per-pixel exposure times. This approach is a true on-focal-plane method which still needs to be implemented on a sensor chip.

High frame rate HDR imaging is a challenging problem, even with state-of-the-art processing units. Thus, many attempts have been made to develop a dedicated hardware processing system for this purpose. Hassan and Carletta [18, 44] proposed an FPGA architecture for Reinhard [38] and Fattal [13] local operators. Even though the proposed implementations concern only the tone mapping operator, the designs require a lot of resources. This originates from the Gaussian pyramid and LUT implementation of the logarithm function [18] and a local Poisson solver [44]. Another FPGA system was implemented by Lapray et al. [25, 26]. They presented several full imaging systems on Virtex-5 FPGA platform as a processing core. The system uses a special HDR monochrome image sensor providing a 10-bit data output.

Apart from FPGA systems, GPU implementations of full HDR systems are also available. Akyüz [3] presented a comparison of CPU and GPU processing pipelines for already acquired bracketed sequences. Furthermore, real-time GPU implementations of different local tone mapping operators can be found in [2, 42].

10.1.3 HDR Video

An FPGA platform derived from the one presented in Chap. 6 is used for the practice of the real-time omnidirectional video system [41]. The assembled prototype is

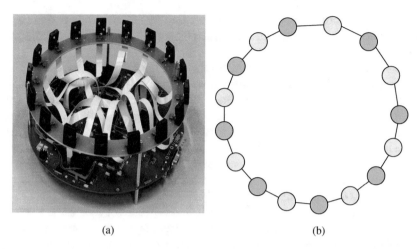

(a)   (b)

Fig. 10.2 (a) The built prototype with processing board at the bottom and the installed camera PCB ring. The diameter of the system is $2r = 30$ cm; (b) the graph representation of the camera arrangement. The *yellow and green circles* represent the cameras with long and short exposure times, respectively. The links between cameras are drawn and each camera can communicate (share pixel values) only with the differently exposed neighbors

shown in Fig. 10.2a. The designed prototype is capable of interfacing maximum 49 cameras to achieve the full hemispherical view. For the purpose of HDR video application, only 16 cameras are placed on a circular PCB ring. The PCB ring in Fig. 10.2a is $2r = 30$ cm in diameter. Low-cost cell-phone VGA cameras, with the minimum FOV of 46°, are placed and operated at 25 fps. The graph representation of the camera connections is given in Fig. 10.2b. Each camera is able to communicate, i.e., share pixel data, with at most two neighboring cameras. Thanks to the inter-FPGA connections, cameras are able to obtain information from a neighboring camera, even if they are not connected to the same FPGA.

The pixel streams coming from the cameras are processed in real-time; hence, HDR video is created as a stack of HDR frames in time domain. Construction of each frame can be divided into two independent processes: (1) construction of HDR composite frame, and (2) tone mapping the composite frame to achieve realistic rendering.

10.1.3.1   HDR Composite Frame

The circular arrangement of the cameras on this prototype allows us to generate the panorama using the Gaussian blending method detailed in Sect. 3.5, simplified to a two-dimensional camera arrangement case. To be able to reproduce the HDR image, the cameras are color calibrated. The camera's response curve is recovered using a set of shots of the same scene with different exposure settings. Three out of twelve taken images are shown in Fig. 10.1. The response curve is recovered by applying the algorithm proposed by Debevec and Malik [10]. Only one camera is color calibrated, as it is assumed that the response curve is identical for all installed cameras.

FOVs of the cameras overlap such that each point in space is observed by at least two cameras [1]. We exploit this property and set the camera exposures to different values. During the camera initialization phase, all cameras are set to the auto-exposure mode. The camera with the longest exposure time, i.e., the one observing a dark region, is taken as a reference. In the following step, half of the cameras are set to the reference exposure t_{ref}, while other half is set to $t_{ref}/4$, such that two cameras with overlapping FOVs have different exposure times. The resulting diagram is shown in Fig. 10.2b, where the yellow and green circles represent cameras with long and short exposure times, respectively.

The calibration data provides yaw, pitch, and roll data for each camera. We are able to determine tessellation of the hemispherical projection surface according to the influence of the cameras, using these Euler angles and the focal length of the camera [16]. Each 3D region in the obtained tessellation denotes a solid angle in which the observed camera has dominant influence, whereas boundaries of these regions are lines of identical influence of two cameras. This tessellation is called the Voronoi diagram [9]. The most influential camera within a single tile is called the *principal camera*.

As the calibration parameters are known, the composite image is constructed by projection of the camera frames onto the hemispherical surface. In order to obtain the HDR radiance map, the color calibration data should be included as suggested in [10]. The pixel values C_i are expressed as:

$$\ln C_i = \frac{\sum_j w(I_{j,i}) \cdot \left(g(I_{j,i}) - \ln t_{ref,j} \right)}{\sum_j w(I_{j,i})} \tag{10.1}$$

$$w(I_{j,i}) = \begin{cases} I_{j,i} - I_{j,min} & , \text{if } I_{j,i} \leq \frac{1}{2}(I_{j,min} + I_{j,max}) \\ I_{j,max} - I_{j,i} & , \text{otherwise} \end{cases} \tag{10.2}$$

where j is the camera index, i is the pixel position, C is the composite image, $I_{j,i}$ represents a set of pixels from contributing cameras, g is the camera response function, and $I_{j,min}$ and $I_{j,max}$ are minimum and maximum pixel intensities in the observed camera frame. The camera response function is recovered using the approach by Debevec [10], and it is shown in Fig. 10.3.

The nature of the HDR imaging is to recover the irradiance using sensors with different exposures. Hence, we constrain the expression (10.1) by evaluating it using only two contributing cameras with mutually different exposures. The second camera is referred to as the *secondary camera*.

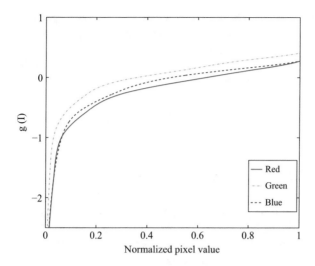

Fig. 10.3 Recovered response function $g(I)$ of a single camera. Three curves correspond to *red*, *green*, and *blue* pixels, as shown in the legend

The calculated piecewise linear weights $w(I_{j,i})$ are included in the Gaussian blending process as:

$$w'(I_{j,i}) = w(I_{j,i}) \cdot \frac{1}{r_j} \cdot e^{-\frac{d_i^2}{2\sigma_d^2}} \qquad (10.3)$$

where the notation is kept identical to (10.1) and (10.2), with r_j as the distance of camera's projection from the observer, and d_i as the pixel distance from the frame center. High standard deviation σ_d increases region of influence of each camera; hence, relative influence of the *principal camera* is reduced. This results in a smoothly blended background and increased ghosting around edges of the objects in the scene. Thus, the standard deviation σ_d is empirically determined for the given camera setup in order to obtain the best image quality.

The result of applying Eqs. (10.1) and (10.3) on the acquired data provides the composite HDR radiance map, which should be tone mapped for realistic display.

10.1.3.2   Tone Mapping

Yoshida et al. [48] made an extensive comparison of the tone mapping operators. The comparison was realized by human subjects grading several aspects of the constructed image, such as contrast, brightness, naturalness, and detail reproduction. One of the best graded techniques in this review was the local operator by Drago et al. [11]. Therefore, this operator will be taken as a base for the development of an

FPGA-suitable operator. Similar to majority of the global operators, this operator uses logarithmic mapping function expressed in (10.4), where displayed luminance L_d is derived from the ratio of world luminance L_w and its maximum L_{max}. The algorithm adapts the mapping function by changing the logarithm base t as a function of the bias parameter b, as shown in (10.5).

$$L_d = \frac{\log_t(L_w + 1)}{\log_{10}(L_{max} + 1)} \tag{10.4}$$

$$t(b) = 2 + 8 \cdot \left(\frac{L_w}{L_{max}}\right)^{\frac{\ln b}{\ln 0.5}} \tag{10.5}$$

Even though this mapping is created for interactive applications, its speed is very slow for video applications. The reported frame rate is below 10 fps, for 720×480 pixels image, without any approximations which decrease the image quality [11]. Calculation of the logarithm values is the most process intensive part [35], whether the algorithm is implemented on CPU or GPU. We have derived an operator suitable for direct hardware implementation which reduces the calculation time.

Drago et al. [11] proposed changing logarithm base and calculating only natural and base-10 logarithms. However, fast logarithm calculations are very resource-demanding, because they require large pre-calculated LUTs. Hence, we approximate the logarithm of the form $\log(1 + x)$ by the Chebyshev polynomials of the first kind $T_i(x)$ [31]. This approximation needs only six integer coefficients to achieve 16-bit precision, which is enough for log-luminance representation in this camera. The Chebyshev approximation can be applied to both natural and base-10 logarithm by only changing the coefficients. The coefficients for the natural logarithm are denoted as $c_e(i)$, while $c_{10}(i)$ are for base-10 in (10.6).

According to [11], the best visually perceived results are obtained for the bias parameter $b \approx 0.85$. Fast calculation of generic power functions, e.g., the one required in (10.5), is not possible. Hence, we fixed the parameter to $b = 0.84$, to relax the hardware implementation, without losing any image quality. The exponent is then 0.25, and the result can be evaluated by two consecutive calculations of the square root. The square root is also approximated by the Chebyshev polynomials. The expanded operator is expressed as:

$$L_d = \frac{\sum\limits_{i=0}^{5} c_e(i) T_i(L_w)}{\sum\limits_{i=0}^{5} c_{10}(i) T_i(L_{max}) \cdot \ln\left(2 + 8\left(\frac{L_w}{L_{max}}\right)^{\frac{1}{4}}\right)} \tag{10.6}$$

The natural logarithm term in the denominator cannot be precisely approximated by Chebyshev polynomials, due the arguments much higher than 1. A suitable approximation of the expression $\ln x$ is a fast convergence form of the Taylor series, which is expressed in (10.7). This expression needs only three non-zero coefficients to achieve a sufficient 16-bit precision, but the argument should be preconditioned as shown:

$$\ln x = 2 \sum_{k=1}^{3} \frac{1}{2k-1} \left(\frac{x-1}{x+1} \right)^{2k-1} \tag{10.7}$$

Equations (10.6) and (10.7) describe the new tone mapping operator suitable for hardware implementation. The set of required mathematical operations is reduced to only addition, multiplication, and division, which are suitable for fast implementation.

10.1.4   FPGA Implementation

10.1.4.1   Local Processing

The processing platform consists of seven slave FPGAs used for local image processing, and each slave unit can be connected to seven cameras, due to I/O pin availability. Local processing is realized on the camera level, utilizing the custom-made Smart Camera Intellectual Property (SCIP) shown in Fig. 10.4. SCIP

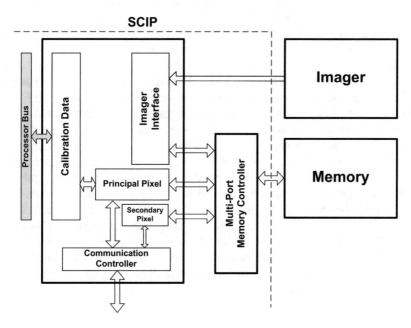

Fig. 10.4 Internal blocks of the smart camera IP used in the slave FPGAs

---

Algorithm 5 Smart camera processing

1: calculate calibration data
2: calculate weights
3: **for all** principal pixels **do**
4:     $p_m := read\_pixel\_from\_memory$
5:     $p_{s,in} := request\_pixel\_from\_secondary\_camera$
6:     $C := \frac{w'_m \cdot p_m}{w'_m + w'_s} + p_{s,in}$
7:     send C to central unit
8: **end for**
9: **for all** secondary pixels **do**
10:     wait for request from principal camera
11:     $p_s := read\_pixel\_from\_memory$
12:     $p_{s,out} := \frac{w'_s \cdot p_s}{w'_m + w'_s}$
13:     send $p_{s,out}$ to principal camera
14: **end for**

---

is instantiated for each camera in the system, and it is in charge of creating a partial HDR composite within camera's FOV.

Responsibilities of each SCIP are three-folded: (1) Acquire pixels from the imager and store them in memory, (2) Evaluate the HDR pixel value where the selected camera is the principal camera, and (3) Provide pixel value to the principal camera, when the selected camera is the secondary camera. The Imager Interface in Fig. 10.4 receives the pixel stream from the camera and stores in the memory. Calibration data block stores information about position of all cameras which are physically close to the observed camera. Thus, SCIP determines the local Voronoi tessellation, and calculates both principal and secondary weights for the camera. The distributed implementation of the algorithm from Sect. 10.1.3.1 is summarized in Algorithm 5.

The principal pixel block is responsible for calculation of the final HDR pixel value. Using the calibration data, the block reads the appropriate pixel from memory, multiplies it with the weight, and requests the weighted pixel from the secondary camera. The secondary camera is not necessarily connected to the same FPGA. Thanks to the communication controller, where camera connection graph is stored, the secondary pixel is obtained [41]. The secondary pixel has already been multiplied by the HDR blending weight in the secondary pixel block, thus only final addition is required. The resulting HDR pixel is further provided to the central unit.

The secondary pixel block operates in the similar fashion. The block waits for the pixel request from the principal camera, reads the pixel from memory, multiplies by the weight, and sends the value back to the principal camera. Both principal and secondary pixel blocks operate concurrently; hence, there is no wait time between principal and secondary pixel processing, which allows very fast calculation time and no loss in the frame rate.

10.1.4.2 Central Processing

The central FPGA acts as a global system controller. The received data comprises sixteen parts of the full HDR panorama, i.e., one part per SCIP. Besides pixel data, SCIPs send information about the correct position in the full-view panorama. Hence, the central unit decodes the position and places the HDR pixel at the appropriate location in the temporary storage memory. When all the pixels belonging to the same frame are received, tone mapping process starts.

The RGB pixel values are read from memory and transformed into the YUV color system, with 16-bit precision per channel. To be in accordance with the previous notation, the values of the pixel luminance channel Y will be denoted by L_w.

The tone mapping implementation consists of two parts: finding the maximum pixel luminance L_{max} and tone mapping curve implementation. Finding L_{max} consists of finding the maximum value in a sequence of the read luminances. L_{max} value is needed for the core tone mapping operation, as shown in (10.6). When HDR video stream is processed, L_{max} is taken from the previous frame, under the assumption that the scene illumination does not vary faster than response time of the HVS. The parameter is updated at the end of each frame.

Figure 10.5 presents the block diagram of the central unit, with emphasized tone mapping block. Chebyshev and Taylor polynomials are evaluated using pipelined implementation of the Horner scheme. The fast Anderson algorithm [31] is used for division implementation.

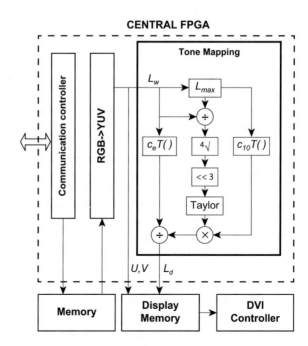

Fig. 10.5 Internal architecture of the central FPGA. Tone mapping block is emphasized as the core processing unit

Taylor series approximation of the logarithm is fast converging only around the center point of the expansion, i.e., $x = 1$ when expansion from (10.7) is used. Even though the luminance value is in the range [0,1], the logarithm argument in the denominator of the tone mapping function (10.6) varies in the range [2,10]. Hence, the argument needs to be brought as close as possible to 1. Since the luminance values are logical vectors (vectors of ones and zeros), the identity (10.8) is used. The number of leading "ones" in the fixed-point representation of the luminance determines the scaling factor y, and the division is implemented as the arithmetical bit-shift-right operation.

$$\ln x = \ln(x/2^y \cdot 2^y)$$
$$\approx \ln(x/2^y) + y \cdot 0.6931$$

$$(10.8)$$

The tone mapped luminance value is combined with the corresponding chrominance components and written into the DVI controller for display.

10.1.5   Results and Discussion

10.1.5.1   Image Quality

The installed cameras have a vertical FOV $= 46°$ and capture 4.9 Mpixels/frame in total. Even though the ratio of vertical and horizontal FOV of the system is 1:8, we experienced that a panoramic strip of size 256×1024 pixels provides enough pixel information, without significant deformation of the objects. This panorama is fitted in the VESA standard XGA frame (768×1024 pixels) and displayed directly on screen using DVI connection. The XGA frame is chosen due to 36 Mb capacity of the dedicated display memory in Fig. 10.5.

In order to quantify the loss in image quality due to applied approximations, the peak signal-to-noise-ratio (PSNR) is calculated for images in the calibration set, whose subset is shown in Fig. 10.1. The HDR image is created and tone mapped in Matlab using approximated and non-approximated calculations. Non-approximated double-precision tone mapped image is taken as the ground truth. Resulting luminance of the approximated tone mapping from Sect. 10.1.3.2 is quantized as a 16-bit value and its PSNR is measured to be 103.61 dB. Thus, luminance of the resulting image does not lose its original 16-bit precision.

Three video screenshots are shown in Fig. 10.6. Figure 10.6a depicts an indoor scene using the automatic exposure mode of the cameras. The measured dynamic range is 1:43. Inside objects are well visible, however, the window region is saturated due to strong light outside of the room. Figure 10.6b shows the same scene rendered using the proposed HDR module. Even though overlap of FOVs is uneven for each camera pair, difference in color tone is not noticeable. Furthermore, the produced image shows details in previously saturated regions, such as the other buildings, while preserving visibility in the darker inside regions. The dynamic

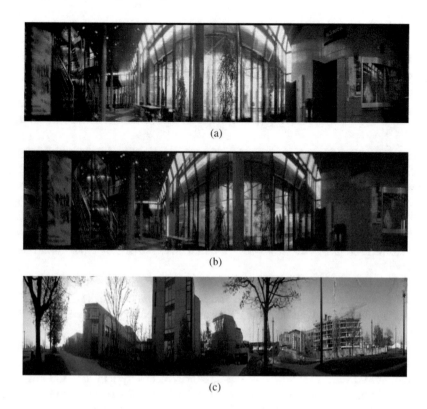

Fig. 10.6 Panoramic HDR reconstruction with a pixel resolution of 256×1024. The cameras were set (**a**) to automatic exposure mode, (**b**) such that two neighboring cameras have different exposure times, one four times shorter, and (**c**) one exposure time eight times shorter, to adapt to bright conditions of outdoor scenery. The blending weights are calculated using $\sigma_d = 300$, to provide sufficient influence of the secondary camera

range of the reconstructed scene is increased to approximately 1:160, which results in 3.72 increase in dynamic range.

The indoor reconstructions suffer from ghosting of near objects due to parallax. The ghosting was expected, because the cameras were calibrated in an environment with no close objects. However, the observed ghosting is different from motion blur, which originates from the difference in exposures. Figure 10.6c shows a rendered HDR outdoor scene, where the closest objects were approximately at 30 m distance. Hence, the edges in this images are significantly sharper than in the indoor environment. The motion blur is not visible around the moving crane or tree branches, thanks to negligible difference in exposure times.

This HDR construction method does not provide as significant increase in dynamic range as some of the other methods, due to the use of only two f-stops. However, up to our knowledge, it is the only system which uses multiple cameras to create and render HDR radiance map simultaneously, and provides real-time HDR video signal at the output.

10.1.5.2   System Performance

The chosen figure of merit for performance of real-time systems is the total processing bandwidth, which best describes the system's capability. The figure of merit is calculated as:

$$\text{BW} = N_{\text{pixels}} \cdot F \cdot \text{BPP} \tag{10.9}$$

where N_{pixels} is the total number of processed pixels, F is the operational frame rate, and BPP is the number of bytes per processed pixel. As Eqs. (10.1)–(10.3) show that all pixels acquired by the presented system are processed, the number of processed pixels is equal to 16 VGA (640×480) frames. The operational frame rate is $F = 25$ fps as input and output frame rates are equal. Each pixel is represented with $BPP = 2$ bytes in RGB format. The conversion to YUV in the central FPGA transforms each pixel into two bytes for luminance, and one byte per chrominance channel.

Comparison of the designed prototype and algorithm implementation with the related systems is given in Tables 10.1 and 10.2. The numbers in the comparison are taken from the original publications if they are published, or calculated by Eq. (10.9) using the available publication data.

Performance comparison shows that the proposed system is superior to the state-of-the-art systems for HDR video construction. The only comparable work is of Slomp and Oliveira [42] with 214 MB/s. However, this system uses the high-end GPU to implement only the tone mapping function. The main reason for high performance of our system is the fully pipelined operation which processes one pixel per clock cycle. Thus, frame rate is linearly dependent on the clock frequency.

Table 10.1 Performance comparison with the related full HDR systems

Type	Full HDR systems			
	This work	[26]	[24]	[15]
Bandwidth (MB/s)	245.7	196.6	112	45
Processing unit	Virtex-5	Virtex-5	GeForce 680	–
Real-time video	Yes	Yes	Yes	Yes

Table 10.2 Performance comparison with the tone mapping systems

Type	Full HDR systems				
	This work	[11]	[2]	[44]	[42]
Bandwidth (MB/s)	245.7	37.8	74	104.85	214
Processing unit	Virtex-5	Fire GL X1	GeForce 8800	Stratix II	GeForce 8800 GTX
Real-time video	Yes	No	No	No	Yes

Table 10.3 Slave FPGA device utilization

Module	Total used	Available
Slices LUTs	63,732	69,120
Slice registers	40,509	69,120
BlockRAM/FIFO	89	128
DSP48Es	61	64

Table 10.4 Central FPGA device utilization

Module	Total used	Available
Slices LUTs	18,376	69,120
Slice registers	17,498	69,120
BlockRAM/FIFO	88	128
DSP48Es	58	64

The utilization summaries of slave and central FPGAs are given in Tables 10.3 and 10.4. The utilization reports are provided for the complete system capable of supporting all 49 cameras.

10.1.6   Conclusion

In this section, we showed how a multi-camera system can be used for real-time HDR video recording. The system produces a real-time HDR video using multiple low-cost cell-phone cameras, i.e., without rather expensive HDR sensors. It is able to simultaneously acquire LDR data, reconstruct an HDR radiance panoramic composite frame, and tone map for realistic display on screen. High system bandwidth and 25 fps frame rate make this prototype an excellent choice for real-time and HDR video applications.

The reconstruction algorithm utilizes the overlap in FOVs of the camera sensors, which are set to different exposure times. We exploit this setup to increase the dynamic range of the captured images and construct an HDR composite image. The HDR image is tone mapped using the fast pipelined global tone mapping algorithm, which was adapted for efficient FPGA implementation.

10.2   Free-View Synthesis Hardware Using Trinocular Disparity Estimation

The recent development of high-quality free viewpoint synthesis algorithms and their implementations allows to realize glasses-free 3D perception. Although many algorithms are developed for this application, the real-time hardware realization of a free viewpoint synthesis for real-world images is challenging due to its high computational load and memory bandwidth requirements. In this section, the first

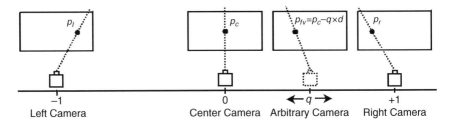

Fig. 10.7 Definition of the three-camera free viewpoint synthesis setup

real-time high-resolution free viewpoint synthesis hardware utilizing the proposed three-camera disparity estimation is presented. The proposed hardware generates high-quality free viewpoint video at 55 fps using a Virtex-7 FPGA at a 1024×768 XGA video resolution for any horizontally aligned arbitrary camera positioned between the leftmost and rightmost physical cameras.

10.2.1 Hardware-Oriented Three-Camera Free Viewpoint Synthesis Algorithm

The simplified classical concept of three-camera horizontally aligned free viewpoint synthesis is illustrated in Fig. 10.7. Assuming that an object is viewed as a single pixel in the x coordinate p_c of the center image, p_l of the left image, and p_r of the right image, then the disparity of a pixel at p_c is $d = p_l - p_c = p_c - p_r$. If the free-view image is generated for an arbitrary camera at location q ($-1 \le q \le 1$, the normalized location of the left camera is at -1, and the right camera is at 1), 3D trigonometry dictates that the view coordinate of the same object in the generated image can be computed as $p_{fv} = p_c - q \times d$. Hence, the pixel at p_c of the center image can be picked and mapped to coordinate p_{fv} of the synthesized image. The free-view synthesis is a complex problem using real-world images considering that the computation of the correct d for every pixel is very challenging especially for object boundaries, low-textured regions, and occluded parts of the images. In addition, the real-time implementation of high-quality free viewpoint synthesis algorithm causes significant challenges due to high computational load and memory bandwidth requirements, especially for high-resolution video.

The proposed hardware-oriented three-camera adaptive weight free viewpoint synthesis (TAW-FVS) algorithm generates the free viewpoint images of any single horizontally aligned arbitrary camera positioned between the leftmost and rightmost cameras. The TAW-FVS algorithm consists of three steps. First, lens distortions and camera misalignments are corrected using camera calibration and rectification. Second, the disparity values of every pixel of the center image are computed using the proposed trinocular adaptive weight disparity estimation algorithm. Third, free views are synthesized by a low computational cost rendering algorithm using the images of the three cameras, the disparity values of the center image, and the position of the arbitrary camera (q).

Internal and external calibration values of the cameras are computed offline using the Open-CV multiple camera calibration toolbox [7]. The Caltech rectification algorithm [6] is used to horizontally align the images captured from the three cameras. The proposed trinocular DE and rendering algorithms are developed to support efficient parallel operations, to consume low hardware resources, and to avoid the requirement of the external memory while providing high-quality free viewpoint synthesis results.

The quality of the free viewpoint images essentially depends on the quality of the disparity estimation. Two-camera DE causes wrong disparity estimation values for the occluded regions. The three-camera DE solves most of the occlusion issues and provides much better DE results, thanks to its double-checking scheme, i.e., using DE results of center-left and center-right image pairs. Moreover, the three-camera system extends the horizontal range of the arbitrary camera location compared to the two-camera version.

In order to obtain high-quality and real-time DE, the trinocular DE algorithm that is presented in Sect. 8.2 is used. The proposed T-AWDE algorithm provides significantly better DE quality than the binocular AWDE by exploiting the fusion of the DE results of the center-left and center-right pairs. During the DE process of every pixel of the center image, the candidate disparities on the right side are searched for the center-left pair, and the candidate disparities on the left side are searched for the center-right pair. The T-AWDE compares the hybrid cost values of the center-left and center-right pairs for every disparity to select the one that exhibits a minimum cost as a disparity value. As an addition to the T-AWDE algorithm, an adaptive penalty is used in the DE step of the TAW-FVS algorithm. An adaptive penalty is utilized depending on the location of the free viewpoint while comparing the cost values computed by center-right and center-left pairs. If $q < 0$, the disparity selection is conducted to select the disparity value assigned by the center-left pair, and if $q \geq 0$, the disparity selection is conducted to select the disparity value computed by the center-right pair by adding $|q| \times 150$ to the matching cost of one pair. Adverse effects of sensitive rectification errors and wrong disparity selections are significantly removed using adaptive penalty. Subsequently, the T-AWDE iteratively smoothens the computed disparity map using the brightness values of the neighboring pixels following the refinement scheme prescribed by AWDE algorithm [8].

The rendering algorithm is applied upon completion of the DE process of the center image. The rendering process involves two steps as presented in Fig. 10.8. First, the arbitrary disparity maps of the arbitrary camera are synthesized by translating the disparity map of the center camera to the location of the arbitrary camera. If $q < 0$, the arbitrary disparity maps of the arbitrary-center and arbitrary-left pairs are generated. If $q \geq 0$, the arbitrary disparity maps of the arbitrary-center and arbitrary-right pairs are generated. In the next step, the free viewpoint is synthesized using the arbitrary disparity maps and the existing camera pictures.

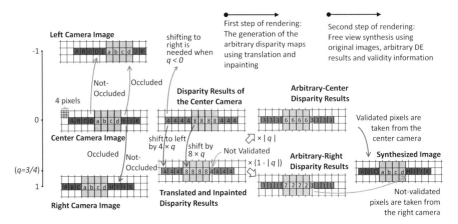

Fig. 10.8 The illustration of free viewpoint synthesis algorithm that utilizes trinocular disparity estimation (Each square grid represents 4×4 pixels. Only one out of four column/row grids is represented to improve the clarity of the representation.)

As presented in Fig. 10.8, more than one pixel of the center camera's disparity image can map to the same pixel of the translated disparity map image. In this case, the higher disparity value belongs to a closer object, thus the higher disparity overwrites the lower disparity. Any pixel that is not mapped during the translation process identifies an occlusion; this location is marked as a *not-validated* pixel location. In order to obtain valid depth values for occluded regions, those pixels are filled by an inpainting method. When the disparity of a pixel is not validated, the disparity data from the last valid value is repeated. This inpainting process is applied from left-to-right when $q < 0$, and right-to-left when $q \geq 0$. The translated and inpainted depth map is then multiplied by $| q |$, $(1 - | q |)$ and again $(1 - | q |)$ to create the arbitrary-center, arbitrary-right, and arbitrary-left disparity values, respectively. Those disparity values and the validity information are used to reconstruct the arbitrary image. The pixels of non-occluded, i.e., *validated*, areas are taken from the center image. The pixels of occluded, i.e., *not-validated*, areas are taken from the left image if $q < 0$, or from the right image if $q \geq 0$.

10.2.2   Real-Time Free Viewpoint Synthesis Hardware

The block diagram of the full system that implements the TAW-FVS algorithm is presented in Fig. 10.9. A Virtex-7 FPGA included in the VC707 Evaluation Board is used to prototype the developed hardware. The cameras are perfectly synchronized using the method explained in Sect. 8.2. All the real-time video processing computations are implemented in hardware. The PC is used as a display, to control the system and to obtain camera calibration parameters. The resulting free views are transferred to the PC using 1 Gb raw Ethernet. A standard 2D TV is

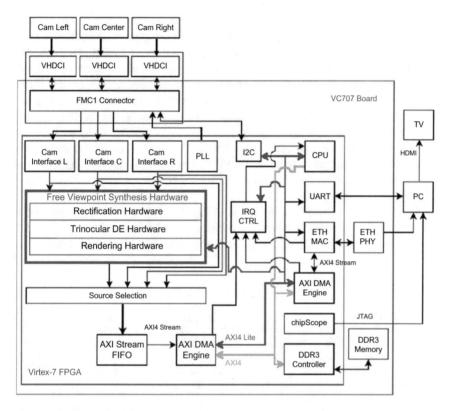

Fig. 10.9 Block diagram of the free viewpoint synthesis system

connected to the PC using HDMI, which offers a better display and emulates a future glass-free 3D TV. A MicroBlaze softcore is used to initialize cameras through I^{2}C, to control Ethernet, and to communicate with the PC. A DDR3 memory is only used for Ethernet buffering. The camera interface, rectification, disparity estimation, and rendering hardware blocks are designed to avoid utilizing DDR3 memory; thus they receive a stream input and provide a stream output. Therefore, the video processing core can be easily converted to a single ASIC. The system is able to send either the 32-bit pixel RGB+Depth video of the center camera, or the RGB video of any physical camera, or the synthesized RGB video of a free view.

The implemented Caltech rectification hardware utilizes 64 on-chip BRAMs for each of the cameras. Each BRAM is used to buffer one row of the image. The rectification hardware processes the images of three cameras in parallel, and synchronously transfers rectified YCbCr images to the disparity estimation module.

The DE hardware of the TAW-FVS system is based on the T-AWDE hardware presented in Sect. 8.2. The T-AWDE hardware is composed of two high-performance and high-quality binocular disparity estimators presented in Sect. 8.1. The outputs of the two estimators are two maps, one associated with the center-left pair of cameras and the other associated with the center-right pair. A fusion block selects which disparity value of these two maps will compose the final three-camera disparity map by comparing the matching costs computed by the two pairs. The fusion block of the DE hardware of TAW-FVS system adds adaptive penalties to the matching costs depending on the position of the arbitrary camera. Finally, three-camera DE hardware iteratively refines the DE values to smoothen the computed disparity map using the brightness values of the neighboring pixels. The disparity estimation hardware buffers the input pixel values using 39 BRAMs for each camera to realize a window-based matching scheme. The 8-bit disparity values and the 24-bit pixels of the left, center, and right cameras are additionally buffered and synchronized using 16 BRAMs for each channel before transferring these data to the rendering hardware.

The rendering hardware is presented in Fig. 10.10. The rendering hardware is blind to the DE process. It receives the synchronized left, center, and right RGB camera images and the disparity image of center camera as its inputs, row by row.

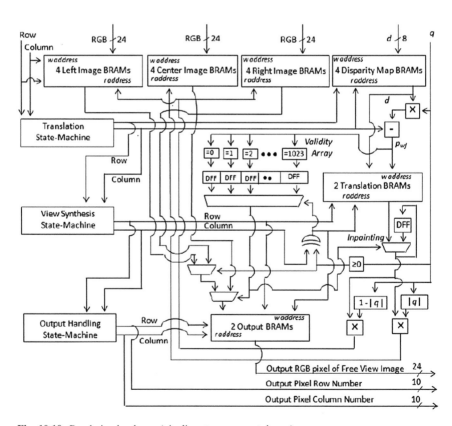

Fig. 10.10 Rendering hardware (pipeline stages are not shown)

The synthesized image is the output. The arbitrary camera location q is dynamically sent from the GUI. q is defined as a 5-bit signed fixed-point value to allow 32 arbitrary locations where "00000" represents the center camera location.

Four BRAMs are used to buffer four rows of each input channel. The rendering process is handled row by row. The least significant bits of the row number are used to select the BRAM. The column number is used as an address. The *translation state machine* generates row numbers and column numbers to read back the disparity map BRAMs pixel by pixel. Concurrently, p_{fv} values are computed using fixed-point multiplication and subtraction hardware, then rounded to the closest integer. p_{fv} is used as write address of the *Translation BRAMs* where the shifted disparity values are buffered. A 2×1024 size flip-flop array is used to buffer the validity bits. The *validity array* keeps track of the *not-validated* pixel locations which require inpainting. Whenever a disparity value is written to the p_{fv} address of the *Translation BRAMs*, the column location p_{fv} is marked as valid by comparing p_{fv} with 1024 possible column addresses using 1024 comparators. The pixels in a row that are not validated remain as *not-validated* pixels since the *validity array* of a row is reset after the process of row synthesis.

The *view synthesis state machine* generates row numbers and column numbers to read back the *Translation BRAMs* and *validity array* pixel by pixel. The reading scan direction of a column is set from 1023 to 0 if $q < 0$, and 0 to 1023 if $q \geq 0$ since the inpainting direction changes according to the arbitrary camera location. The *validity bit* is used as a select signal of a multiplexer. If the pixel is not validated, the disparity value previously stored in the flip-flop replaces it, i.e., paints the read disparity value. The translated and inpainted disparity value is multiplied by $|q|$ and $(1 - |q|)$ to compute the arbitrary disparity values of the arbitrary-center, arbitrary-left, and arbitrary-right camera pairs. Arbitrary disparity values are used as a read address for the input channel BRAMs. The selection of the source pixel is handled by multiplexers. If the pixel is validated, the pixel of the center image is used as the synthesized pixel. If $q < 0$ and the pixel is not validated, the pixel of the left image is used as the synthesized pixel. If $q \geq 0$ and the pixel is not validated, the pixel of the right image is used as the synthesized image pixel. Synthesized image pixels are buffered in two output BRAMs. The *output handling state machine* reads back the output BRAMs pixel by pixel, and sends outputs of the rendering hardware as the final output of the free viewpoint synthesis hardware.

10.2.3 Implementation Results

The proposed real-time three-camera free viewpoint synthesis hardware is implemented using Verilog HDL, and verified using Modelsim 10.1d. The Verilog RTL models are mapped to a Virtex-7 XC7VX485T FPGA comprising 607k look-up tables (LUT), 303k DFFs, and 1030 BRAMs. The rectification hardware consumes

13% of the LUTs, 6% of the DFF, and 19% of the BRAM resources of the FPGA. The trinocular DE hardware consumes 25% of the LUTs, 11% of the DFF, and 16% of the BRAM resources of the FPGA. The rendering hardware consumes 1% of the LUTs, 1% of the DFF, and 2% of the BRAM resources of the FPGA.

The system is functionally verified in real-time. The speed performance of the free viewpoint synthesis hardware core is limited by the working frequency and parallelization of trinocular DE hardware. The hardware core operates at 175 MHz after place and route. The trinocular DE hardware and thus the complete TAW-FVS system can process up to 55 fps at a 1024×768 XGA video resolution for a 128 pixels disparity range. Although this speed performance is verified using Chipscope, the current display output of the system is 18 fps due to bandwidth limitation of raw Ethernet output. The 55 fps performance of the hardware will be fully exploitable using USB3 or HDMI in the future.

Real-time snapshots captured by the proposed system are presented in Fig. 10.11. While capturing multiple consecutive snapshots, models stand stable in front of the system. The horizontal locations of some specific locations are marked offline with dashed lines in original left, center, and right images in Fig. 10.11a, b, c, respectively. The center-left, trinocular, and center-right DE results provided for the center camera image are presented in Fig. 10.11d, e, f, respectively. The trinocular DE solves most of the occlusion and wrong estimation errors by the fusion of the DE results of the two pairs which is essential to synthesize high-quality free view images. The synthesized free viewpoint images of two artificial viewpoints $q = (-0.5)$ and $q = 0.5$ are presented in Fig. 10.11g, h, respectively. The efficiency of the presented system is evidenced by horizontal pixel coordinates of the marked location in the synthesized image. Moreover, correct 3D appearance and occlusion effects in the synthesized free viewpoint images can be identified by observing the background objects near the models in Fig. 10.11.

10.3 Software-Based Real-Time Applications of Disparity Estimation

The software running on a PC client delivers three fundamental tasks: the capture and display of the video output of the FPGA, allowing user to send commands to the system, and developing software-based real-time applications. Several software applications are developed using the disparity estimation hardware. The implemented software applications are: speed and distance measurement, depth-based image thresholding, head-hands-shoulders tracking, virtual mouse using hand tracking, face tracking integrated with free viewpoint synthesis, and stereoscopic 3D display. The implementation details of these applications are presented in this section.

High-resolution visual results of the depth estimation system are demonstrated in the videos below:

https://zenodo.org/record/16544
http://dx.doi.org/10.5281/zenodo.16544

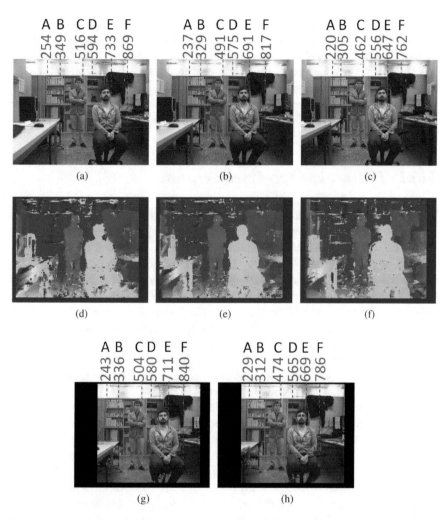

Fig. 10.11 Real-time snapshots captured by the proposed system. (**a**) Left image, (**b**) center image, (**c**) right image, (**d**) IR-AWDE for center-left, (**e**) T-AWDE, (**f**) IR-AWDE for center-right, (**g**) synthesized free viewpoint image for an arbitrary camera located at $q = (-0.5)$, (**h**) synthesized free viewpoint image for an arbitrary camera located at $q = (0.5)$ (images best viewed in high resolution from the pdf files)

The GUI of the proposed depth estimation system, the visual results of video processing cores, and the software-based application operated in PC are demonstrated in the video files "DepthEstimationSystemVideoExample1.mov" and "DepthEstimationSystemVideoExample2.mp4." The videos evidence that the proposed depth estimation system provides accurate depth estimation results, and it can be used for a wide range of 3D-based video processing applications.

The user is able to select the display format and software application through the GUI. Several software applications are developed using the disparity estimation system. The software applications are implemented to evidence that the high-quality RGB+D outputs of the proposed system can be used to develop advanced 3D-based video processing applications in real-time, though improving the theory of the software applications itself is not targeted. Future possible improvements of the software applications and their prospective advanced utilizations are explained.[1]

10.3.1 Distance and Speed Measurements

The system sends a disparity map to the PC. The depth of the pixel or the distance of the object from the camera can be calculated by using this disparity value by the GUI. The baseline chosen for the initial setup is $B = 10$ cm and the focal length of the lenses is $f = 6$ mm. Since the unit of the disparity is the number of the pixels, the f value should be also represented using the number of pixels. The f value with the unit of pixels is obtained as *941* from the camera calibration. The f and baseline values are written to the GUI by the user, a single time at the system power-on. If different type of optics are used or if the distance between the cameras are changed, the user is required to change the new fixed values in the Main tab of the GUI. The GUI is able to convert the disparity d to a distance z for every pixel. Since the computational load of disparity to depth conversion is very low, and it can easily run in real-time, the conversion is not implemented in hardware.

Since the distance can be computed for every pixel, computing the speed of the pixel into the direction z is also straightforward. The GUI computes the speed of the pixel in terms of cm/s using the frame rate and the change of distance z of a pixel, from image to image. The expression of the computation of speed (spd cm/s) is provided in Eq. (10.10), where z_n and z_{n+1} identify the distance of the pixel with the unit of cm in two consecutive images n and $n+1$, respectively. fps represents the instantaneous frame rate of the camera. Currently, the speed and distance value can be numerically visualized for every clicked pixel from the GUI. The distance image is visualized in the monitor for all pixels. The high-resolution speed image can be visualized in the future, which can be useful to measure and display the speed of the vehicles and sportsmen.

$$\text{spd} = (z_{n+1} - z_n) \times \text{fps} \qquad (10.10)$$

---

[1]The software applications head–hands–shoulders tracking, virtual mouse using hand tracking, and face tracking integrated with free viewpoint synthesis are developed in EPFL, as a collaboration with Youngjoo Seo, a PhD candidate at KAIST, Korea.

10.3.2   Depth-Based Image Thresholding

A depth-based thresholding application is developed to be used as an alternative method of the green-wall technique used in the cinematography. In cinematography, green walls are used to replace the background with another image or video. Since the proposed system provides high-quality RGB+D outputs for every pixel, a similar technique can be implemented without using green walls.

In this application, first, the snapshot image of the static background is taken. The threshold disparity value is provided by the GUI. If measured disparity (d) of any RGB pixel is higher than the given threshold, it is regarded as the frontground and the RGB value is displayed. Otherwise, the pixel of the background image that is captured at system initialization is displayed. Therefore, if the object is far enough, it disappears from the display, and it is displayed if it is close enough.

In the current application, the system operates with the background image of the same environment. The background image can be changed with the picture of another environment in the future. In addition, instead of static background, a video of the another environment can be used as a background in the future. Thereby, any person or object can integrated in a video of another environment in real-time, which can be useful for the journalists for the presentation of live-TV news and sport events.

10.3.3   Head, Hands, and Shoulders Tracking Using Skeltrack Software

The head–hands–shoulders tracking application is implemented using the open-source Skeltrack software [19]. Skeltrack is originally implemented to benefit from the depth estimation results of Kinect. It utilizes a data-driven approach presented in [4] for real-time human skeleton detection. It does not use any database of pre-recorded body positions. In replacement of Kinect's depth result, the input to the software is obtained from the depth results of the proposed system. The Skeltrack software efficiently performs in real-time to track the head, hands, and shoulders. The skeleton detection can be used to play computer games. In addition, the skeleton detection can be integrated with human pose estimation and gesture recognition techniques in the future. Since the presented system can operate outdoors, the human behavior can be detected outdoors which can be used in surveillance applications or in sports events. Occlusion is an important problem in efficient human pose estimation. Several proposed systems can be used to monitor the same environment without interference. Therefore, the user can be surrounded by multiple of proposed depth estimation systems to provide efficient real-time pose estimation results without occlusion problem.

10.3.4 Virtual Mouse Using Hand Tracking

The virtual mouse application is implemented by tracking hands and fingers using the algorithm presented in [45]. Convex and concave points in the hand are found using the depth image. The concave locations present the finger tips. Less than two concave points on the hand are a sign of a closed palm, i.e., a fist pose. The first pose of the hand is utilized to identify clicking. If the hand is open, the user is able to unclick and freely move the mouse. The virtual mouse application can be used to play computer games or implement touch-less device control. A similar finger tracking and identification technique can be used to implement touch-less keyboard application in the future.

10.3.5 Face Tracking Integrated with Free Viewpoint Synthesis

The free viewpoint synthesis requires to receive an arbitrary camera location q from the GUI. Two different methods are implemented to dynamically change the arbitrary camera location. In the first method, the cursor on the GUI is utilized to switch between different viewpoints in real-time. In the second method, the location of the user with respect to the monitor is used to dynamically change the q values. The face tracking algorithm presented in [20] is implemented in the GUI to detect the location of the user. The RGB camera of the PC is used for the face tracking. The integration of the face tracking in to the GUI to interact with the free viewpoint synthesis hardware allows the user to freely move its observation point between the cameras. This implementation can be used to allow the users to watch sports events in real-time from any viewpoint that they select, in the future.

10.3.6 Stereoscopic 3D Display

The system is able to switch to stereoscopic 3D display mode to visualize green and blue values obtained from one camera and the red value obtained from the other camera. Currently the system can be used to visualize 3D using color anaglyph-based 3D glasses.

The 3D visual quality can be increased using active-glasses-based 3D display in the future. In addition, since the system is able to provide a disparity result for every pixel, it can be used for implementing advanced augmented-reality applications. Any 3D object can be artificially placed at any distance, and the user can interact with these arbitrary objects.

References

1. Afshari H, Popovic V, Tasci T, Schmid A, Leblebici Y (2012) A spherical multi-camera system with real-time omnidirectional video acquisition capability. IEEE Trans Consum Electron 58(4):1110–1118. doi:10.1109/TCE.2012.6414975
2. Akil M, Grandpierre T, Perroton L (2012) Real-time dynamic tone-mapping operator on GPU. J Real-Time Image Process 7(3):165–172. doi:10.1007/s11554-011-0196-7
3. Akyüz AO (2012) High dynamic range imaging pipeline on the GPU. J Real-Time Image Process 10:273–287. doi:10.1007/s11554-012-0270-9
4. Baak A, Müller M, Bharaj G, Seidel HP, Theobalt C (2013) A data-driven approach for real-time full body pose reconstruction from a depth camera. In: Consumer depth cameras for computer vision. Springer, London, pp 71–98
5. Bloch C (2013) The HDRI handbook 2.0: high dynamic range imaging for photographers and CG Artists. Rocky Nook, Santa Barbara
6. Bouguet JY (2004) Camera calibration toolbox for matlab. http://www.vision.caltech.edu/bouguetj/ [Online]
7. Bradski G, Kaehler A (2008) Learning OpenCV: computer vision with the OpenCV library. O'Reilly Media, Incorporated, Sebastopol
8. Chander G, Markham BL, Helder DL (2009) Summary of current radiometric calibration coefficients for Landsat MSS, TM, ETM+, and EO-1 ALI sensors. Remote Sens Environ 113(5):893–903
9. de Berg M, van Kreveld M, Overmars M, Schwarzkopf O (2000) Computational geometry: algorithms and applications, 2nd edn. Springer, Berlin
10. Debevec PE, Malik J (1997) Recovering high dynamic range radiance maps from photographs. In: ACM SIGGRAPH 97, New York, NY, pp 369–378
11. Drago F, Myszkowski K, Annen T, Chiba N (2003) Adaptive logarithmic mapping for displaying high contrast scenes. Comput Graphics Forum 22(3):419–426. doi:10.1111/1467-8659.00689
12. Durand F, Dorsey J (2002) Fast bilateral filtering for the display of high-dynamic-range images. ACM Trans Graph 21(3):257–266. doi:10.1145/566654.566574
13. Fattal R, Lischinski D, Werman M (2002) Gradient domain high dynamic range compression. ACM Trans Graph 21(3):249–256. doi:10.1145/566654.566573
14. Granados M, Ajdin B, Wand M, Theobalt C, Seidel HP, Lensch HPA (2010) Optimal HDR reconstruction with linear digital cameras. In: Proceedings of 23rd IEEE conference on computer vision and pattern recognition (CVPR), pp 215–222
15. Gupta M, Iso D, Nayar S (2013) Fibonacci exposure bracketing for high dynamic range imaging. In: IEEE international conference on computer vision (ICCV)
16. Hartley RI, Zisserman A (2004) Multiple view geometry in computer vision, 2nd edn. Cambridge University Press, Cambridge
17. Hasinoff SW, Durand F, Freeman WT (2010) Noise-optimal capture for high dynamic range photography. In: Proceedings of 23rd IEEE conference on computer vision and pattern recognition (CVPR), pp 553–560
18. Hassan F, Carletta JE (2007) An FPGA-based architecture for a local tone-mapping operator. J Real-Time Image Process 2(4):293–308. doi:10.1007/s11554-007-0056-7
19. Jocha J (2012) Skeltrack : a free software library for skeleton tracking. https://github.com/joaquimrocha/Skeltrack [Online]
20. Jun B, Kim D (2012) Robust face detection using local gradient patterns and evidence accumulation. Pattern Recogn 45(9):3304–3316
21. Jungmann JH, MacAleese L, Visser J, Vrakking MJJ, Heeren RMA (2011) High dynamic range bio-molecular ion microscopy with the timepix detector. Anal Chem 83(20):7888–7894
22. Kalantari NK, Shechtman E, Barnes C, Darabi S, Goldman DB, Sen P (2013) Patch-based high dynamic range video. In: ACM transactions on graphics (TOG) (Proceedings of SIGGRAPH Asia 2013), vol 32(6), p 202

23. Kang SB, Uyttendaele M, Winder S, Szeliski R (2003) High dynamic range video. ACM Trans Graph 22(3):319–325. doi:10.1145/882262.882270
24. Kronander J, Gustavson S, Bonnet G, Unger J (2013) Unified HDR reconstruction from raw CFA data. In: Proceedings of IEEE international conference on computational photography
25. Lapray PJ, Heyrman B, Rosse M, Ginhac D (2012) HDR-ARtiSt: high dynamic range advanced real-time imaging system. In: IEEE international symposium on circuits and systems, pp 1428–1431. doi:10.1109/ISCAS.2012.6271513
26. Lapray PJ, Heyrman B, Ginhac D (2014) HDR-ARtiSt: an adaptive real-time smart camera for high dynamic range imaging. J Real-Time Image Process, 1–16. doi:10.1007/s11554-013-0393-7
27. Mann S, Picard RW (1995) On being 'undigital' with digital cameras: extending dynamic range by combining differently exposed pictures. In: Proceedings of IS&T, pp 442–448
28. Mantiuk R, Daly S, Kerofsky L (2008) Display adaptive tone mapping. ACM Trans Graph 27(3):68:1–68:10
29. Martinez-Sanchez A, Fernandez C, Navarro PJ, Iborra A (2011) A novel method to increase LinLog CMOS sensors' performance in high dynamic range scenarios. Sensors 11(9):8412–8429. doi:10.3390/s110908412
30. Mertens T, Kautz J, Van Reeth F (2007) Exposure fusion. In: Pacific conference on computer graphics and applications, pp 382–390. doi:10.1109/PG.2007.17
31. Meyer-Baese U (2007) Digital signal processing with field programmable gate arrays, 3rd edn. Springer, Berlin
32. Mitsunaga T, Nayar S (1999) Radiometric self calibration. In: IEEE conference on computer vision and pattern recognition (CVPR), vol 1, pp 374–380
33. Pattanaik SN, Tumblin J, Yee H, Greenberg DP (2000) Time-dependent visual adaptation for fast realistic image display. In: ACM SIGGRAPH 00, New York, NY, pp 47–54. doi:10.1145/344779.344810
34. Pattanaik SN, Reinhard E, Ward G, Debevec PE (2005) High dynamic range imaging - acquisition, display, and image-based lighting. Morgan Kaufmann, Burlington
35. Popovic V, Pignat E, Leblebici Y (2014) Performance optimization and FPGA implementation of real-time tone mapping. IEEE Trans Circuits Syst II: Express Briefs 61(10):803–807. doi:10.1109/TCSII.2014.2345306
36. Portz T, Zhang L, Jiang H (2013) Random coded sampling for high-speed HDR video. In: IEEE international conference on computational photography (ICCP). doi:10.1109/ICCPhot.2013.6528308
37. Ramachandra V, Zwicker M, Nguyen T (2008) HDR imaging from differently exposed multiview videos. In: IEEE 3DTV conference: the true vision-capture, transmission and display of 3D video, pp 85–88
38. Reinhard E, Stark M, Shirley P, Ferwerda J (2002) Photographic tone reproduction for digital images. ACM Trans Graph 21(3):267–276. doi:10.1145/566654.566575
39. Robertson MA, Borman S, Stevenson RL (2003) Estimation-theoretic approach to dynamic range enhancement using multiple exposures. J Electron Imaging 12(2):219–228
40. Saleem A, Beghdadi A, Boashash B (2012) Image fusion-based contrast enhancement. EURASIP J Image Video Process 2012(10):1–17. doi:10.1186/1687-5281-2012-10
41. Seyid K, Popovic V, Cogal O, Akin A, Afshari H, Schmid A, Leblebici Y (2015) A real-time multiaperture omnidirectional visual sensor based on an interconnected network of smart cameras. IEEE Trans Circuits Syst Video Technol 25(2):314–324. doi:10.1109/TCSVT.2014.2355713
42. Slomp M, Oliveira MM (2008) Real-time photographic local tone reproduction using summed-area tables. In: Computer graphics international, Istanbul, pp 82–91
43. Tocci MD, Kiser C, Tocci N, Sen P (2011) A versatile HDR video production system. ACM Trans Graph 30(4):41:1–41:10. doi:10.1145/2010324.1964936
44. Vytla L, Hassan F, Carletta JE (2013) A real-time implementation of gradient domain high dynamic range compression using a local Poisson solver. J Real-Time Image Process 8(2):153–167. doi:10.1007/s11554-011-0198-5

45. Wah Ng C, Ranganath S (2002) Real-time gesture recognition system and application. Image Vision Comput 20(13):993–1007
46. Ward G (1991) Real pixels. In: Graphics gems II. Academic, San Diego, CA, pp 80–83
47. Ward G, Rushmeier H, Piatko C (1997) A visibility matching tone reproduction operator for high dynamic range scenes. IEEE Trans Vis Comput Graphics 3(4):291–306. doi:10.1109/2945.646233
48. Yoshida A, Blanz V, Myszkowski K, Seidel HP (2005) Perceptual evaluation of tone mapping operators with real-world scenes. In: SPIE human vision & electronic imaging X, pp 192–203. doi:10.1117/12.587782

Chapter 11
Conclusion

In this book, we presented a new approach to designing multi-camera systems, which consists of merging the image sensors and the processing units into a single system. Unlike state-of-the-art systems that use cameras purely for image acquisition, and a setup of PCs for its offline processing, the two presented camera systems use FPGAs for real-time image processing. We presented the complete design flow including the hardware design choices, algorithm development for panorama construction, FPGA implementation, user interface options, and finally, real-time tests.

The first part of this book regards development of a hardware-suitable real-time panorama construction algorithm. Most of the currently used stitching and blending algorithms are developed and optimized for running on a CPU or a GPU. Their direct hardware implementation is either not fast enough for real-time operation, or does not provide acceptable image quality. We presented a novel image blending algorithm called Gaussian blending, which builds up on a well-known alpha blending. The implementation of Gaussian blending on our camera system provides seamless panoramic image, and suppresses appearance of ghosts. Furthermore, Gaussian blending can be implemented in real-time for any number of cameras in the system without loss of image quality. The disadvantages and limitations of the Gaussian blending are that the variance of the used Gaussian curve has to be determined manually, and that the ghost suppression around objects sometimes results in slightly blurry edges. To overcome this problem, we introduced a probabilistic approach to panorama generation, and we have shown how it can be implemented in a real-time system.

The second part of the book is the full design of five different multi-camera systems. The first system is a miniaturized multi-camera system called Panoptic. The system is built using fifteen commercially available low-cost cell-phone cameras placed on a hemispherical dome. The real-time image processing is performed on a custom-made Virtex-5 FPGA board. The panorama construction is implemented using the Gaussian blending and provides good quality results.

V. Popovic et al., *Design and Implementation of Real-Time Multi-Sensor Vision Systems*, DOI 10.1007/978-3-319-59057-8_11

The second system is a 24-camera system prototype targeted for endoscopy applications. Furthermore, this is the first multi-camera system with a distributed built-in illumination capability featuring fiber optic technology. The third system shows that the image processing does not need to be centralized in one unit, but that it can be distributed in a network of smart camera nodes with limited local processing, and central frame assembly. Furthermore, with this system we demonstrated the ability of the camera to create the environment for virtual reality goggles Oculus Rift, which gives an immersive experience to the user. Up to our knowledge, this is the first camera-goggles system that can also be used where real-time telepresence is required. The fourth design is the design of a high-resolution multi-camera system, GigaEye II. This system also works in real-time, thanks to the novel, distributed panorama construction. The system is designed as a multi-layered processing system, with each processing unit dedicated to reconstructing the partial FOV. Using such approach, it is possible to process and support very large data rates. The system can currently process all 320 Mpixels that are captured each frame. The system is designed to be modular, and the increase in resolution can be achieved by placement of the new cameras and addition of one FPGA processing board per four cameras. The fifth system demonstrates the capability to integrate depth estimation algorithms into a binocular or trinocular camera system, and achieve real-time depth estimation, which was not possible before with the regular passive camera stereo pair and CPU processing.

Finally, the last part of this book presents possible applications of the designed multi-camera systems.

11.1 Future Prospects

The future of the real-time computational imaging systems can be discussed in the following groups:

- **Algorithms**: The proposed Gaussian blending for Panoptic and multi-band blending for GigaEye II provide high-quality results, but are not optimal solutions. Recently, several new methods were developed that make use of the optimization algorithms. Implementing an optimization in real-time on FPGA is a very challenging and interesting problem, which has not been solved up to now. Solving it would definitely make a significant impact in real-time computational imaging, as well as many other research fields.
- **Miniaturization**: Panoptic camera utilized compact camera modules to reduce the system size. However, the processing PCB and FPGA are large with respect to the dome where the cameras are placed. The design of an ASIC to replace the FPGA, reduce the number of external components, and reduce the PCB size is the next logical step. Furthermore, the design of an ASIC with certain level of programmability can lead to commercialization of Panoptic as a consumer product.

- **True gigapixel video**: The finalization of the true gigapixel real-time omnidirectional sensor. The "upgrade" to higher, gigapixel resolutions has already been envisioned, and its realization is only a matter of need for such a high-resolution system.
- **New applications**: The new applications can be proposed for any of these real-time multi-camera systems, such as 3D cinematography, video super-resolution, embedding virtual reality content in the scene (for gaming).
- **Displays**: A major drawback for high-resolution camera systems is the inability to display the full image, and demonstrate its full potential in real-time. Hopefully, the fast-growing advancements in the camera technology will be followed by very high-resolution displays in the near future.

Printed in the United States
By Bookmasters